T0276049

SpringerBriefs in Applied Sciences and Technology

Computational Intelligence

About this Series

The series "Studies in Computational Intelligence" (SCI) publishes new developments and advances in the various areas of computational intelligence—quickly and with a high quality. The intent is to cover the theory, applications, and design methods of computational intelligence, as embedded in the fields of engineering, computer science, physics and life sciences, as well as the methodologies behind them. The series contains monographs, lecture notes and edited volumes in computational intelligence spanning the areas of neural networks, connectionist systems, genetic algorithms, evolutionary computation, artificial intelligence, cellular automata, self-organizing systems, soft computing, fuzzy systems, and hybrid intelligent systems. Of particular value to both the contributors and the readership are the short publication timeframe and the world-wide distribution, which enable both wide and rapid dissemination of research output.

More information about this series at http://www.springer.com/series/10618

António Daniel Silva · Rui Ferreira Neves
Nuno Horta

Portfolio Optimization Using Fundamental Indicators Based on Multi-Objective EA

António Daniel Silva
Instituto Superior Técnico
Instituto de Telecomunicações
Lisbon
Portugal

Nuno Horta
Instituto Superior Técnico
Instituto de Telecomunicações
Lisbon
Portugal

Rui Ferreira Neves
Instituto Superior Técnico
Instituto de Telecomunicações
Lisbon
Portugal

ISSN 2191-530X ISSN 2191-5318 (electronic)
SpringerBriefs in Applied Sciences and Technology
ISBN 978-3-319-29390-5 ISBN 978-3-319-29392-9 (eBook)
DOI 10.1007/978-3-319-29392-9

Library of Congress Control Number: 2016930544

Printed on acid-free paper

This Springer imprint is published by SpringerNature
The registered company is Springer International Publishing AG Switzerland

To Griselda Ventura Porfirio

António Daniel Silva

To Susana and Tiago

Rui Ferreira Neves

To Carla, João and Tiago

Nuno Horta

Preface

The investment problem of achieving higher returns and at the same time reducing the risk is a very complex and non-deterministic problem. This brought the attention of computer science in general and artificial intelligence in particular to address the problem of financial investing. Scientists and engineers are using artificial intelligence to create algorithms to manage portfolios with very low human intervention to take decisions. Some of the techniques used are genetic algorithms, genetic programming, neural networks, simulated annealing, and tabu search. These algorithms are used to predict price movements, find graphical patterns, select the best ones from a range of assets, for doing arbitrage and hedging. More recently, they have been applied to high-frequency trading by analyzing the supply and demand in the market. The goal to achieve is a higher return than the market with a lower risk, and using stocks instead of another type of financial instruments such as options, futures, and Forex, it is probability the best choice to take to constitute a portfolio. The advantage in selecting this type of assets is related to the diversity of available stocks, the lower correlations between them, and the lower volatility of some stocks.

This work presents a new approach to portfolio composition in the stock market. It incorporates a fundamental approach using financial ratios and technical indicators with a multi-objective evolutionary algorithms to choose the portfolio composition with two objectives, the return and the risk. Here, two different chromosomes structures are used for representing different investment models with real constraints equivalent to the ones faced by managers of mutual funds, hedge funds, and pension funds. In order to validate the present solution, two case studies are considered using the SP&500 from June 2010 until the end of 2012. The simulations demonstrate that stock selection based on financial ratios is a combination that can be used to choose the best companies in operational terms, obtaining returns above the market average with low variances in their returns. In this case, the optimizer found stocks with high returns on investment in a conjunction with a high rate of growth of the net income and a high profit margin. To obtain stocks with a high valuation potential it is necessary to choose companies with a lower or average market capitalization, low PER, high rates of revenue growth, and high operating leverage.

Chapter 1 describes the problematic addressed by this book, the portfolio optimization using EA.

Chapter 2 presents background information and reviews the existing literature, discussing existing approaches that are relevant for developing this project.

Chapter 3 describes the components needed to evaluate companies suc as the financial analysis and the meaning of the important item in each statement. Additionally, a brief explanation is given on the technical analysis approach used to evaluate the market and to enter or exit a position.

In Chap. 4 a detailed description on the multi-objective approach used in this work is presented.

In Chap. 5 the experiences performed in single and multi-objective optimization are explained, and the results are discussed.

Chapter 6 presents the conclusions and the future work.

António Daniel Silva
Rui Ferreira Neves
Nuno Horta

Contents

Abbreviations

Optimization and Computer Engineering Related

ACO	Ant Colony Optimization
AI	Artificial Intelligence
ATI	Automatic Trading Systems
EA	Evolutionary Algorithm
GA	Genetic Algorithm
MO	Multi-objective Optimization
PS	Pareto Set
PSO	Particle Swarm Optimization
SA	Simulated Annealing
TS	Tabu Search

Investment Related

ACP	Average Collection Period
APP	Average Payment Period
B&H	Buy-and-hold
CAPM	Capital Asset Price Model
CR	Current Ratio
DCF	Discount Cash Flow
DER	Debt Equity Ratio
DY	Dividend Yield
EBIT	Earnings Before Interests and Taxes
EMH	Efficient Market Hypothesis
EPS	Earnings Per Share
FA	Fundamental Analysis
GDP	Gross Domestic Product
HFT	High Frequency Trading

HSI	Hang Seng Index
MA	Moving Average
MC	Market Capitalization
NYSE	New York Stock Exchange
PBV	Price Book Value
PCF	Price Cash Flow
PER	Price Earnings Ratio
POR	Pay Out Ratio
QR	Quick Ratio
ROA	Return On Assets
ROE	Return On Equity
S&P 500	Standard and Poor's 500 stock index
TA	Technical Analysis

List of Figures

List of Tables

Chapter 1
Introduction

Abstract This chapter describes the problematic addressed by this book, the portfolio optimization using EA, the main goals for the work, and the document's structure.

The stock exchange is the physical place where equity shares of companies listed in that exchange trade using manual or advanced automatic trading system (ATS). The largest and best-known exchange is the New York Stock Exchange (NYSE) where thousands of US companies are registered and traded. The evolution of the Internet and globalization allow trading stocks through Internet using a broker, with a commission and low slippage.[1] The possibility to know the transaction price of the stocks and volumes in real time[2] is fundamental to trade any common stock of the NYSE or in other exchanges, because it allows the trader to perform market analysis and know the execution price of his orders.

The advances in computer science, like the increase of the CPU's processing power and the memory capacities and hardware, led to a large increase in the available data on financial markets. This allows researchers to apply a new set of tools, evolutionary algorithms (EA), artificial intelligence (AI), fuzzy logic, and machine learning, to solve a variety of problems such as portfolio optimization and prediction of price movements of the financial instruments [1].

Investors use these applications to create trading programs to execute trades without human intervention, in different time periods that can be annual, weekly (swing trading), daily (day trading), and of less than 1 h (high-frequency trading). There are other types of programs developed to explore arbitrage opportunities and for portfolio selection. These systems are backtested in historical or current markets to check the possible performance that can be achieved, and improve the trading system before they are used. The main approaches to invest in the stock market are

[1]The difference between the expected price of the trade and real executed price.
[2]With a low delay in time.

A.D. Silva et al., *Portfolio Optimization Using Fundamental Indicators Based on Multi-Objective EA*, SpringerBriefs in Computational Intelligence,
DOI 10.1007/978-3-319-29392-9_1

supported by either technical analysis, fundamental analysis, or the random walk theory. Technical analysis studies the market patterns and demand and supply of stocks shares. Fundamental analysis studies a stock, from the economic and financial viewpoint, where it tries to define if a company is undervalued in the market in relation to intrinsic value, calculated by the projections of future earnings of the company. Random walk theory defends that the market discounts all future developments so that the investor cannot expect to outperform the market [2].

In the US stock market (S&P 500, Dow Jones, Nasdaq 100), participants annually trade trillions of dollars, they are short-term traders, scalpers, speculators, companies doing hedge to reduce their risk, and value investors [3]. Short-term traders invest in a period lower than a week, basing their investment decisions in technical analysis. Scalpers are traders that use technical analysis to gain a little spread between bid and ask values, in trades with a period of time inferior to 5 min. Speculators use time frames higher than a week, normally months, to trade using technical and fundamental analysis. Value investors use fundamental analysis to evaluate a company and buy a share below its intrinsic value.

This book relies mostly on the value investor ideas from Warren Buffett, the famous investor who sustains that a diversified portfolio of common stocks, each of them purchased below its intrinsic value, will generate a return above the general market.

To create a portfolio, a manager needs to take a set of decisions to solve the portfolio optimization problem and distribute his capital in a number of assets to maximize the expected return and minimize the risk. To solve this problem it is necessary to consider the correlation of the assets, budget constraints, and the investor's preference. These types of problems are complex and practically impossible to solve by exact methods, and researchers use heuristics like local search (LS), tabu search (TS), simulated annealing (SA), particle swarm optimization (PSA), ant colony optimization (ACO), and EA to solve them.

EA are based on the process of biologic evolution, the solutions of the problem are encoded in data structures, chromosomes, and tested using a fitness function. The previous best solutions are used to generate new solutions, by repeating this process the algorithm is capable of being close to the global optimal solution of the problem.

1.1 Motivation

Competition among investors is high and many techniques used to invest in financial instruments do not produce outstanding results. This results in a motivation and the challenge to implement an uncommon technique that generates better results than the stock market Index S&P 500. Another reason is to test the efficient market hypothesis theory (stock price reflects all information available and this implies that it is impossible to beat the index with an active management approach[3]), and try to find and explore some inefficiency in the market valuations.

One important aspect of any investment system is its adjustment where the performance obtained in training and real test are similar. Financial markets are uncertain and dynamic, change very quickly, so the goal is to develop a robust system to invest that can outperform the index, and a strategy of buy and hold, in the most adverse environments that a stock market can offer.

1.2 Approach

The selection of the best companies to invest is probably the most important factor for the success of any investor of common stocks. In this research it will be investigated the effectiveness of computer intelligent systems to invest, using the value investing approach. EA will be used to select a business to invest that have the principles of Warren Buffet, using fundamental analysis of the financial statements (income statement, balance sheet, cash flow statement), and ratio analysis.

1.3 Problems Description

The use of a fundamental approach based on financial statements analysis, to select the stocks that will have in the future a better performance in the market, is a difficult task because most of the fundamental information is discounted in the valuation done by the market.

Using EAs in a training environment, for later use in real market situation, led to the most difficult problem to deal, the adjustment of the solutions, namely the performance that the system will achieve in other markets and conditions.

It is expected that a robust investing system will be independent of the market conditions and will have a low variance on the different performance measures. To guarantee a good fitting it is essential to use a large sample of data in the training, and avoid the over-fitting[4] effect.

Another problem to deal with, when using some EA as search mechanism, is the classical selection process that can lead to the algorithm to get trapped in a local optimum and not reach the global optimum, which causes great loss of performance in the trading system.

[3]Portfolio management strategy where the goal is to outperform an investment benchmark.

[4]A modeling error which occurs when a function is too closely fit to a limited set of data points.

1.4 Document Structure

This book is structured as presented as follows:

- Chapter 2 presents the necessary theory and methodologies used to perform this investigation.
- Chapter 3 presents the architecture of the system, and the methods used.
- Chapter 4 presents the methodologies used in the EA and the developed algorithms.
- Chapter 5 summarizes the experiences and the obtained results.
- Chapter 6 outlines the conclusions of the work and suggests some topics for future work.
- The appendix describes the recorded investments considering the two selected strategies.

References

1. M. Kaucic, Portfolio management using artificial trading systems based on technical analysis. Genet. Algorithms Appl. (2012)
2. B. Malkiel, A random walk down wall street (1999)
3. C. Faith, Way of the turtle (McGraw-Hill, New York, 2007)

Chapter 2
Literature Review

Abstract This chapter presents the background information and reviews the existing literature relevant to the development of this project. In the first part of the chapter, a brief description of the existing investment approaches is presented. Particularly, in Sect. 2.1, the "*Value Investing*" and the related instruments and methods will be described in depth. In Sect. 2.2 the technical analysis and the corresponding strategies and tools to analyze the market are presented. A formal definition of multi-objective optimization (MO) problems and their concepts are given in Sect. 2.3. A brief description of evolutionary computation and evolutionary algorithms is presented, in Sects. 2.4 and 2.5, before the review of the existing literature on MOEA's for portfolio management, in Sects. 2.6 and 2.7.

2.1 Value Investing

Value investing is a comprehensive investment philosophy to perform in-depth fundamental analysis to limit risk, resist the crowd psychology,[1] and achieve long-term investment results. It is also the practice of purchasing securities or assets for less than its intrinsic value [1].

Value investors seek stocks of companies that trade below their estimations of intrinsic value in the market, they act based on the theory that markets overreact to good and bad news resulting in stock price movements that do not correspond with the company's long-term fundamentals. Next, the analysis of financial statements, the use of economic indicators, the fundamental ratios, the method of investment of Warren Buffett, and the approaches to evaluate the stock price are presented.

[1]Members of the crowd often adapt and act to the expectations of the surrounding culture and modify individual traits in order to identify with the crowd.

A.D. Silva et al., *Portfolio Optimization Using Fundamental Indicators Based on Multi-Objective EA*, SpringerBriefs in Computational Intelligence,
DOI 10.1007/978-3-319-29392-9_2

2.1.1 *Fundamental Analysis*

A company's stock can be undervalued because of a disappointing earnings report, a restructuring of the company, a lawsuit, if the company fails its strategic or financial objectives, bear market tendency of the general market that provokes short-term trend in the price of the stock.

Fundamental analysis entails rigorous evaluation of the company's fundamentals as industry growth, the capacity of the competitors, the future global economic factors for its goods and service, and macroeconomic conditions that affect the company.

The work of fundamental analyst to evaluate a security entails the study of the financial statements and industrial and ratio analysis. The analyst needs to be always looking to find information about corporate actions, like restructuring, spin-offs,[2] merger and acquisitions, to get a better estimation of the true value of the company.

2.1.2 *Financial Statements*

Financial statements are reports issued by companies to demonstrate their financial performance, normally trimestral or annual. The objective of the financial statements is to provide information about the financial position, performance, and changes in financial position of the company. These reports are composed of an income statement, balance sheet, and cash flow statement and are used by the value investors to do quantitative analysis on the company.

(a) **Income Statement**

Income statement measures the performance of the company for a specific period of time, shows how much money the company operations generated (Revenues/sales), the operating expenses occurred during the period, the operating profit (EBIT), the financial costs (interest and taxes), and the net income [2].

Revenue or sales represents the amount of money a company earns through the sale of a good or service during a specific time period, sometimes managers break down revenue by business segment or geography. The best way for the company to improve its profitability is by increasing its revenues [3].

- **Competitive advantage in the income statement**
  A company has a competitive advantage when they have some power in the market that other rivals do not have; normally these are monopolistic or oligopoly companies. They can price their products much higher than their marginal costs, obtaining higher operating margins [4]. In a case of increase of costs

[2]The creation of an independent company, through the sale of part of the company.

of production, it can be transferred to the final price for the consumer, if this is not possible the company does not have the competitive advantage and the increase in costs will generate lower profit margins.
Companies with gross profit margin above 40 % are more likely to have a competitive advantage and companies with low profit margins are probably in highly competitive industries [3].
In case of a competitive advantage as the result of some technological advancement, there is always a threat that a newer technology will replace it. Companies with business in the area of technology spend a lot on Research and Development, like 30 % of its gross profit or more, which means that probably they are inserted in very competitive industry and maybe not a good company to invest [3].

- **Interest Coverage**
Interest expenses represent the money out of the business to pay the interests of the debts that the company has in the balance sheet, it is a financial cost that depends on the level of leverage of the company. If the company has a low *Times Interest Earned ratio (TIE)* it is highly risky to invest, because an unexpected event that reduces the *EBIT* of the company can put it in position of not fulfilling the payment of its obligations.

$$\text{TIE} = \frac{\text{EBIT}}{\text{Anual Interest Expense}} \tag{2.1}$$

(b) **Balance Sheet**

Balance sheet is the instantaneous report of the financial condition of the company on a particular date and is generated normally at the end of a trimester, or year. This report states what the company owns and how it is financed, it is divided into three parts, assets, liabilities, and total equity with an accounting relationship given by Eq. 2.2 [2, 5].

$$\text{Assets} = \text{Total Equity} + \text{Liabilities} \tag{2.2}$$

Using the report format under U.S. generally accepted accounting principles (GAAP) the assets are registered in the balance sheet in the order of liquidity, in the top are the more liquid assets (the current assets), in the bottom the less liquid assets.

The part of assets are divided into current assets (assets that can be converted in cash in a short period of time), and noncurrent assets (assests that cannot be converted into money so quickly).

The first assets to appear are the cash and short-term investments, if exists a record of accumulation of money year after years, it is a good signal for the investor, in terms of profitability and stability of the company operations [3].

Liabilities are divided into two parts, the short-term liabilities, money that company owed and need to pay in 1 year or less time, this include account payables, accrued expensive, and short-term debt, the other part is the long-term liabilities that come due in more than 1 year.

In industries that have a fiercely competitive environment, the companies need to constantly upgrade their manufacturing facilities to stay competitive. The acquisition of assets creates ongoing expenses that are recorded in property/plant/equipment and according to US GAAP it is reported in the statement at the cost of acquisition less accumulated depreciation.

The recent year's assets acquisitions demonstrate the necessity of upgrade of the operations of the company and this helps to differentiate the companies with competitive advantage from the others.

Companies with higher profit margins have a ratio of property/plant/equipment to total debt higher than the others, meaning lower operations leverage. The high value of fixed assets represents a barrier to competitors to enter in is sector, because they need to do great investment [3].

Companies with durable competitive advantage require little or no long-term debt to maintain their operations, it is necessary to analyze more than one year of the financial statements to see if the operations work with low levels of debt. It may occur that a company generates so much profits that it decides to take advantage of tax shield generated by use of a higher level of debt; in such cases it is normal that net earnings can pay off long-term debt in 3 or 4 years.

Common stocks represent ownership of the company, give the right to elect a board of directors and to receive dividends. Preferred stock is other class of equity that does not give the right to vote, but has the right to receive a dividend before common stocks dividends [2].

Retained earnings are profits that the company retained to invest in the operations of the company, or in others investments considered profitable. The retention ratio defines the percentage of profits that is not distributed as dividends and is retained [2].

(c) **Cash Flow Statement**

Since the accrual method allows credit sales to be booked as revenue in the income statement, it is necessary to keep separate track of the actual cash that flows in and out of business, this is shown in the cash flow statement.

This statement reports only how much money enters (cash inflow) from ongoing operations and external investment sources, and how much money goes out of the company (cash outflow) to pay the business activities, investments, and financial expenses, during the period.

Companies with durable competitive advantage working in their favor generate a great quantity of cash inflow from operations that can be applied by the managers to buy back the company shares reducing the number of outstanding shares. These actions are taken because the managers prefer to finance the company with debt, because it is cheaper; second, they try to manipulate the price of stock in the market, or the manager board thinks the company is undervalued in the market and by repurchasing their own shares can increase the wealth of the shareholders. This can be noted in the statement in the section of *cash from investing activities,* by analyzing the previous years reports [2, 3].

2.1.3 Fundamental Indicators to Use in Ratio Analysis

Ratio analysis is used to conduct quantitative analysis of the information of financial statements, it uses the ratios calculated from the current year and compared with ratios of previous periods to check the performance of the company and to select the best company in an industry. Next, in Tables 2.1, 2.2, 2.3, 2.4, and 2.5, some of the financial ratios used by investors are presented.

Table 2.1 Profitability ratios

Indicator	Profitability ratios	Description
Return on equity (ROE)	$\frac{\text{Net income}}{\text{Shareholder's Equity}}$	ROE measures corporation's profitability, by revealing how much profit a company generates with the money shareholders have invested
Return on assets (ROA)	$\frac{\text{Net income}}{\text{Total Assets}}$	ROA measures efficiency of management using the assets to generate earnings

Table 2.2 Liquidity ratios

Indicator	Liquidity ratios	Description
Current ratio (CR)	$\frac{\text{Current Assets}}{\text{Current Liabilities}}$	Measures the company's ability to pay short-term obligations
Quick ratio (QR)	$\frac{\text{Current Assets} - \text{Inventories}}{\text{Current Liabilities}}$	Measures the company's ability to pay short-term obligations with its most liquid assets

Table 2.3 Leverage ratios

Indicator	Leverage ratios	Description
Debt equity ratio (DER)	$\frac{\text{Total Liabilities}}{\text{Shareholders Equity}}$	Measures of financial leverage
Debt ratio (DR)	$\frac{\text{Total Liabilities}}{\text{Total Assets}}$	Indicates the leverage of the company along with the potential risks the company faces in terms of its debt load
Times interest earned ratio (TIER)	$\frac{\text{EBIT}}{\text{Anual Interest Expense}}$	A metric used to measure a company's ability to meet its debt obligations

Table 2.4 Efficiency ratios

Indicator	Efficiency ratios	Description
Average payment period (APP)	$\frac{\text{Account Payable}}{\text{Sales}} \times 12$	Average period taken by the company to pay to its creditors
Average collection period (ACP)	$\frac{\text{Account Receivables}}{\text{Sales}} \times 12$	Average period for the company receive its payments from its clients

Table 2.5 Market value ratios

Indicator	Market value ratios	Description
Earnings per share (EPS)	$\frac{\text{Net income} - \text{Dividends on preferred stock}}{\text{Average Oustanding Shares}}$	The portion of a company's profit, by each outstanding share of common stock
Payout ratio (POR)	$\frac{\text{Dividends per Share}}{\text{Earnings per Share}}$	This ratio shows the percentage of the earnings that are distributed to shareholders as dividends
Dividend yield (DY)	$\frac{\text{Annual Dividends per Share}}{\text{Share Price}}$	A financial ratio that shows how much a company pays out in dividends each year, relative to its share price

2.1.4 Economic Indicators

Economic indicators are separated into macroeconomic indicators to effectuate macroeconomic analysis to forecast the cycle of the economy, and the industry indicators to analyze the industries.

(d) **Macroeconomic Indicators**

Macroeconomic factors have a huge impact on the general market tendency and influence the economic fundamentals of the companies. It is important to know how to interpret and analyze the indicators presented next.

- **The gross domestic product (GDP)** is a measure of the economy's total production of goods and services. Growing in GDP indicates an expansion of economy with the opportunity for companies to increase their sales and services.
- **Balance of trade** is the difference between a country's imports and its exports. A positive balance attracts more capital (external and internal) for investments. Companies in these countries are more competitive internationally, and are more capable of increasing their operations and sales.
- **Inflation rate** is the rate at which the general level of prices rises. High rates are associated with economies with demand for goods and services where there is outstripping productive capacity, which leads to upward pressure on prices [6].
- **Sentiment of consumers and producers** has important impact in the performance of the economy, an optimistic sentiment in relation to the future causes a behavior to consume more products and services in the consumers and the

producers are willing take more risks through investments and increases of production.
- **Interest rate** is thc cost of money; high rates decrease the present value (PV) of future cash flows, meaning that companies and investors will see decreasing attractiveness of their investments.
- **Monetary expansion**, depending if it is anticipated by the financial markets, can lead to an increase of stock price, because that expansion means lower interest rates for some time and higher output of the economy [7].

I. **Industrial indicators**

The performance of any company depends in great part to the economic future of its industry; industrial indicators are used to perform a sector analysis to have a clear picture of the actual state and future of the sector [6].

- **The number of customers** is important to evaluate the risk of the company, if total revenue was achieved by few customers or some millions is an important factor to evaluate the risk. A loss of one client when the enterprise has few clients or large portion of the revenue is generated by a couple of them is a situation riskier than when there are a large number of clients, and each of them without meaningful contribution for the total revenue.
- **Market share** gives some important information about the company and the business. A company that has more than 70 % of the market suggests that probably it has a competitive advantage like a barrier to enter into the market, or economies of scale.
- **Industrial growth** is a reference for estimating the growth rate of any company that composes the group. In a fast growing industry, companies will follow this growth and the best competitors can have higher growth rate than the industry.
- **The number of competitors** gives information about the competitive environment of the industry, when there exist lower number of barriers for entry and a large number of competing firms create a difficult operating environment for generating profits. With a low competition environment in the sector the companies can use the ability of pricing power and increase their profits or pass increase of production cost to the clients.

2.1.5 Categories of Companies

Lynch and Rothchild define six general categories of companies depending on the growth rate of the company, capitalization size, and its economic behavior [8].

1. **Slow growers** are large and aging companies that are expected to grow slightly faster than the GDP, with time every fast-growing industry becomes a slow growth industry and most of the companies in the sector lose momentum too and became slow growers. The best strategy to apply for investing in this case is

to purchase the shares with the objective to win a dividend, the aspects that investor needed to consider is the payout ratio percentage (lower the better), and see in the financial statements if the company has a good record of pay dividend and check its growth rate.

2. **The Stalwarts** are big companies (multibillion dollars), where their earnings growth is faster than the slow growers. The stock of this type of company purchased at fair price probably will generate 30–50 % gain in a good year. During recessions and hard times (bear market), normally these types of companies tend to perform better than the general market, this happens because they have a durable competitive advantage that in recession times allows the company to increase its earnings.
3. **The fast growers** are small companies, new enterprises that grow more than 20 % a year. They do not necessarily belong to fast-growing industries to have high rates of growth, but can expand in a slow-growing industry by taking the market share of its competitors. The share price of these companies represents the investments with the most potential for valorization, but they too have a higher risk of down fold and a higher volatility in the stock prices than the other types of enterprises.
4. **The cyclical** companies are inside industries that expand and contract with a high correlation with the GDP growth. The sales and earnings grow faster than in other types of companies in periods of economic expansion, but in a scenario of economic recession these companies can go bankrupt or need to pass many years for the industry to recover.
5. **Turnarounds** are companies that pass hard times caused by bad news about the future of the business, poor financial condition that can mean bankruptcy, or a scandal of corruption that affect the price of the stock. When the situation or factor that caused the devaluation of the share price is solved, a quick recovery of the price is probable.
6. **Asset play** is any company that holds a valuable asset that is worth more than what is recorded in the balance sheet, or it is unknown the existence of it by the market in its valuation of the stock.

2.1.6 Warren Buffett Method

The Warren Buffett method consists in investing exceptional companies, with a durable competitive advantage, normally an economic monopoly or oligopoly, where the increase of the price of its products, leads to growth of operating profit. The most important factor for him is the durability of the competitive advantage, if it lasts for a long period of time the tendency to the business and stock price is to increase in the long run.

Warren Buffett invests only in business inside what he defines by "*circle of competence,*" those are companies inside industries that an investor understand and had a profound knowledge [9]. The circle can be amplified and improved with research and experience, this mean that an investor can increase the number of hypotheses to invest and diversifies his portfolio.

He looks for companies with a consistent earning history and with favorable long-term prospects. The business needs to have a good return of equity ratio, low debt, high profit margins, and be managed by honest people for him to invest. After finding a company with these characteristics he calculates the intrinsic value of the shares by discounting the PV of his estimations of future cash flows of the investment, and only purchases a share below its calculations more a margin of safety.

The characteristics of companies are divided into three parts, business, management, and financial.

1. Business characteristics:
   - The business needs to be simple and understandable for the estimation of future cash flows with a high degree of confidence.
   - Business with a consistent operating history gives the guarantee that competitive advantage is durable.
   - Sell a unique product or service.
   - Companies with bargaining power in its purchases (occurs when a company that purchases a large fraction of an industry's output can demand price concessions) or seller power (company has a monolithic control on type of product and can demand higher prices) [6].
   - Company with low need of innovation and new products (meaning a Research and Development costs lower).
2. Management characteristics:
   - Management needs to be honest and rational to allocate the capital for providing returns above the cost of capital.
   - Management should be candid with the shareholders by disclosing all the information relevant to them.
3. Financial tenets:
   - Low debt in balance sheet.
   - Consistent growth in earnings.
   - High ROE, the companies can give a higher return for each dollar invested.

The method of investing by Warren Buffett defends to hold the investment as much time as possible, but sometimes it sells a winning position due to some motives. The first reason to sell is when a better investment is found, and need capital for that investment. Second, when it is almost certain the company will lose its durable competitive advantage. The last reason is when the stock price reaches value that is far beyond the estimated theoretical value the future of the business and the economy is considered [3].

2.1.7 Approaches to Equity Valuation

In this section, two groups of general approaches for evaluating companies are presented, respectively, the discount cash flow techniques, where the value of the stock is estimated based on the PV of the future cash flows, and the relative valuation techniques, that measure the level of overvaluation or undervaluation of a stock based upon its current price relative to earnings, cash flow, book value, and sales.

(e) **Discount Cash Flow Model (DCF)**

DCF valuation models recognize that common stock represents the ownership of a business and the value must be related to the future returns of owners.

Blanchard defines intrinsic value of stocks as the PV of the future expected cash flows, by estimating the total cash flows that are likely to occur in the life of the business and discounting it at an appropriate rate [7].

$$\mathrm{PV} = \sum_{i=1}^{n} \frac{\text{Cash Flow}_i}{(1+R)^i} \tag{2.3}$$

R Discount rate
i Year to discount.

Financial analysts have developed different versions of the DCF, the *dividend discount model* (DDM), the *free cash flow discount model*, and the *residual income model* [2].

Next, the three cases using the DDM to evaluate equity, depending on the rate of growth of dividends, are presented.

I. **Case with a zero growth rate**
In a case that the company pays a constant dividend and the investor holds the stock forever, the value of equity is given by Eq. 2.4

$$P_0 = \sum_{i=1}^{n} \frac{\mathrm{Div}_i}{(1+R)^i} = \frac{\mathrm{Div}_1}{R} \tag{2.4}$$

R Discount rate
Div_i Dividend year i.

II. **Case with a constant growth rate**
Considering that the company will keep a dividend growth rate constant during his life, the value of the equity is determined by Eq. 2.5.

$$P_0 = \frac{\text{Div}_1}{R - g} \tag{2.5}$$

g Growth rate of dividends.

III. **Case with differential growth rate**
This case is more complex, and occurs in the real world. For determining the value of a share the financial analyst needs to do a prevision of the future growth rates of dividends applied to Eq. 2.6. More correct are the previsions of the analyst as he can take better advantage of the inefficiencies in market, due to valuation errors from other investors.

$$P_0 = \frac{\text{Div}_1}{1+R} + \frac{\text{Div}_1 \times (1+g_1)}{(1+R)^2} + \cdots + \frac{\text{Div}_{n-1} \times (1+g_n)}{1+R^n} \tag{2.6}$$

(f) **Relative Valuation Techniques**

These techniques estimate the economic value of a company by comparing it with similar companies and past values. The relative ratios compare the stock price to the financial and economic information of the company.

The PER, price to cash flow (P/CF), price to book value (P/BV), and price to sales (P/sales) will be discussed.

I. **Price earnings ratio**
A valuation ratio of a company's current share price compared to its per-share earnings. It is a way to estimate the value by determining how much dollars are necessary to pay for one of the earnings.

$$\text{PER} = \frac{\text{Share Price}}{\text{EPS}} \tag{2.7}$$

II. **The price/cash flow**
The price over cash flow is a ratio similar to PER, which measures the firm's financial health, the definition uses the cash flow to calculate the ratio, because of that this ratio do not have the effect of depreciation and noncash factors, and represent the price to pay for the money entering in the company.

$$\frac{P}{\text{CF}} = \frac{\text{Share Price}}{\text{Cash flow per share}} \tag{2.8}$$

III. **Price/book value ratio**
The price over book value ratio measures the valuation done by the market to the assets of the company, the premium pay over the book value represents the valuation of the capacity to growth and the ROE.

$$\frac{P}{B} = \frac{\text{Share Price}}{\text{Total Assets} - \text{Intangible Assets} - \text{Liabilities}} \tag{2.9}$$

IV. **Price/sales ratio**
The price over sales ratio evaluates a company in terms of the sales, it is an important ratio because the sales growth is a request to the company's growth. This ratio is used by investors because it is difficult to manipulate the sales in the statement, and give a more exact relative evaluation.

$$\frac{P}{S} = \frac{\text{Share Price}}{\text{Sales}} \tag{2.10}$$

2.2 Technical Analysis

Technical analysis is a method of evaluating securities by analyzing the market activity, the past prices, and transactions volume. The technical analyst uses a set of tools like indicators and chart patterns to determine the possible future movement of the market. The balance of supply and demand is the factor that changes the price and volume of the securities.

This theory ignores totally the fundamental analysis and the intrinsic value of the securities; it is based on the philosophy that the change in the price and volume discounts the fundamental factors and information available [10].

Another important belief is that history tends to repeat itself in terms of price movements. This is justified by the human mentality, the investors and traders tend to have the same reactions and actions in similar conditions of the market. The Chartist technicians analyze the actual chart pattern that the market is forming and compare with past patterns to predict the future market movement. The Dow theory is other form of technical analysis that defines the markets as three trend movements, and each trend is composed of a phase of accumulation of stocks, absorption phase, and distribution phase. The trend to be real needs to be confirmed by increase of volume and they finish with a definitive signal of end.

2.2.1 *Trends in the Market*

The prices fluctuate constantly in the market, but sometimes tend to describe oscillations with a tendency corresponding to a bull market in case of the tendency for price growth, when the trend decreases in price it is called bear market. The Dow theory describes the market as three forces simultaneously that affect the stock price, described below.

(g) **Primary Market Trend**

The main tendency or primary trend is the price tendency with higher time frame, usually lasting several months or years. Defined by economic fundamentals of the company and economy, it starts when the market recognizes the fundamentals and moves in concordance with them, creating a price trend.

(h) **Secondary Market Trend**

The secondary trend corresponds to market fluctuations against the primary trend with a lower period of time. These trends are also often called price correction when the market evolves too quickly relative to their fundamentals and performs a secondary tendency to correct the price.

(i) **Tertiary or Minor Trends**

Minor trends are daily fluctuations of little importance, normally called noise.

2.2.2 *Strategies to Invest*

Technical traders define two principal strategies to invest the trend following and band trading. Next, these strategies will be explained.

(j) **Trend Following**

In this approach sometimes the stock price moves in one direction for a long period of time, generating a large price movement [11].

Trend followers enter long in the market when the price break a reference value and enter short with the inverse conditions. The strategy to be successful needs to have good exit rules to limit its losses and the trader needs to be disciplined to stay with the movement for a long period of time to generate the expected return.

(k) **Band Trading**

Band trading strategies are used when the investor assumes that the market is in a range bound, the trading concept is the opposite of trend following, where the band traders buy at prices close of reference low and sell at a reference high [12].

2.2.3 *Types of Markets*

There are different types of volatility in the market and they can be very dynamic and changing, sometimes very quickly. Faith defines four types of volatility in any financial instrumented trade [12].

(l) **In a range stable and quiet**

Prices tend to stay within a relatively small range with little movement up or down outside the range, as can be seen in Fig. 2.1. The strategy to explore this market is the band trading, but sometimes the volatility is so low that is better to stay out of the market, waiting for a break of the range, and using a trend following strategy.

(m) **In a Range Stable and Volatile**

Prices tend to stay within a big range that is formed in time horizon of weeks, where inside the range has great volatility, as demonstrated in Fig. 2.2. Band trading is the correct strategy to apply when this type of market is identified.

(n) **Trending and Quiet**

Prices move in one direction, with retracements with low volatility in opposite direction, Fig. 2.3. The trend following is the appropriate strategy to use in this case.

- **Trending and Volatile**

Trending and volatile happen when there are large changes in price in one direction, with occasional significant short-term reversals of direction, Fig. 2.4. This case is similar to the one before, but it needs to adapt the trend following strategy to the higher volatility of the correction.

2.2.4 *Designing a Strategy to Invest*

After developing the approach to invest there is the need to create a system to implement the strategy. The building blocks of any strategy are the trading costs, markets to invest, market timing, protection capital, and exit of a wining position.

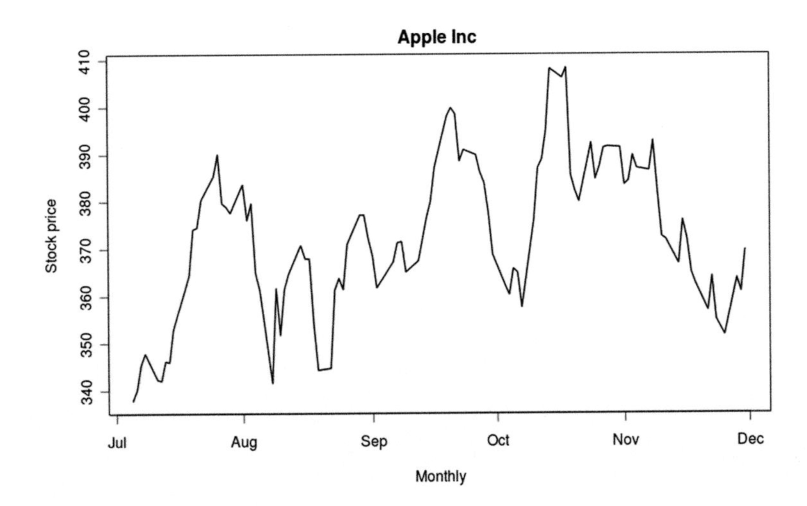

Fig. 2.1 Market with a stable and quiet volatility

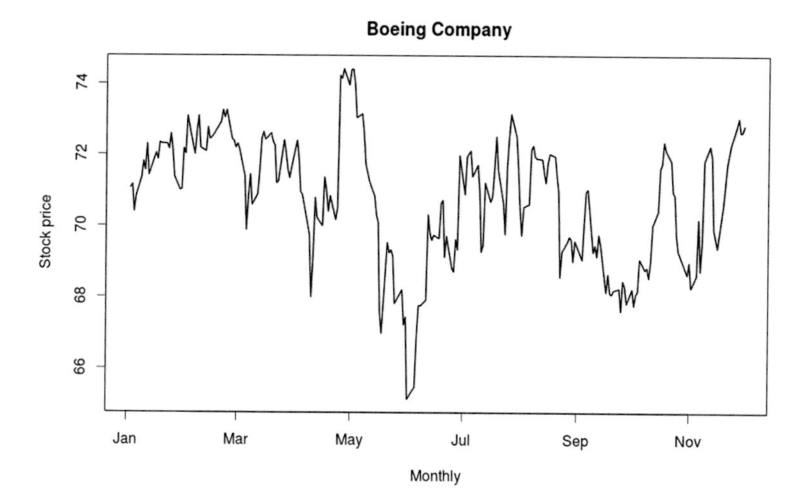

Fig. 2.2 Market stable and volatile

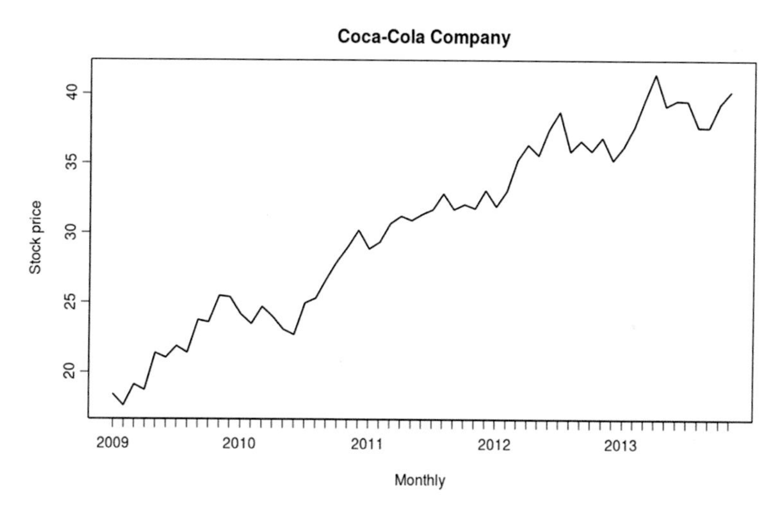

Fig. 2.3 Trending and quiet

(p) Trading Costs

Trading costs are expenses incurred when buying or selling securities, they include the commissions and spreads charged by the brokers for doing the transaction for their clients.

The costs of doing business are important because they reduce the net return of the investments and need to be considered in the investment decisions. High transaction costs and high frequency of trading mean a great percentage of money

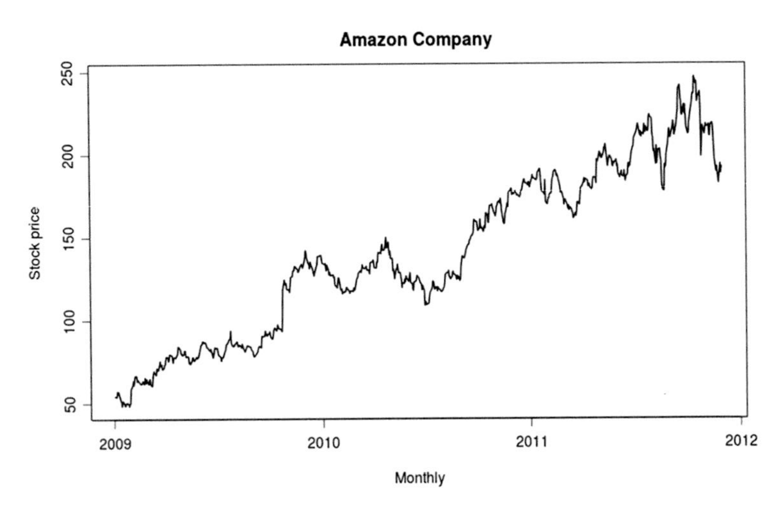

Fig. 2.4 Trending with high volatility

paid in commissions, this can be the difference between being profitable or not in the long run. Investor should avoid possible high commissions and stocks with low liquidity, because these are the stocks with higher spreads.

(q) **Select Markets to Invest**

The first decision of any investor is to select the markets to trade, considering the available budget, possible cost in transactions, the knowledge it has about the different products, the access to information, and its experience.

(r) **Entry or Market Timing**

This is the exact price and the market conditions that need to be present for entering in position of a common stock, the objective is defined a set of rules that give the entry signal to improve the timing of buying and increase the reliability of the system [11].

For *Peter Lynch* the right time to enter is during collapses, free falls, and correction of the retracement of the price, due to lowest price of the securities [8].

(s) **Stops Loss and Protection of Capital**

Stop loss are predetermined policies that reduce a portfolio exposure, it is a component of the system to get out of a losing position, not allowing one or more investments to continue to losing money and protect the remaining capital available to continue the investing activity.

(t) **Exit of a Winning Position**

In any system it is important to define the exit conditions of a winning position, for profits maximization, to perform this are incorporated a set of rules in the system, for selling a position. Technical traders normally sell their positions, when one or more of the next conditions are achieved:

- Sell signal given by a technical indicator
- A retracement of the market in the opposite direction
- When an amount of time pass from the time of entry
- A type of chart pattern done by the market
- When a predefined amount of profit is achieved.

2.3 Introduction to Multi-objective Optimization

Multi-objective optimization (MO) is a method used to solve real-world problems that involve simultaneously several incommensurable and often competing objectives. Normally there is not only a single optimal solution, but a set of optimal solutions.

Multi-objective optimization is the problem of finding the best solutions to optimizing two or more objectives that are in conflict, subjected to certain constrains [13].

The objective function (OF) is an equation to be optimized (maximized or minimized), composed of variables with need to respect certain constraints. Mathematical algorithms are used to optimize the OF, in single-objective optimization there is only one OF to optimize, and in multi-objective there are two or more. The nomenclature used for these functions is $f_i(x)$.

The search space Ω is composed of n parameters, the decision variables. The solutions found for the optimization of the OF are called vector solution or decision vector, they are composed of a set of values each one corresponding to one decision variable. The process of optimization corresponds to varying the value of the decision variables to search in the space the optimal solutions to the OF.

$$x = (x_1, x_2, x_3, \ldots, x_n) \tag{2.11}$$

2.3.1 *Multi-objective Description*

The MO problem can be described as for each point in the space (Ω) defined by n decision variables, which have a vector function $F(x)$ composed of m functions $f_i(x)$. The goal is to find all the solution vectors that solve the problem, considering the restrictions imposed. The mathematical formulation to the problem is described in Eq. 2.12:

$$\begin{aligned} &F(x) = [f_1(x), \ldots, f_m(x)], \quad m \text{ number of OF} \\ &g_i(x) \le b_i, \quad \text{for } i = 1, \ldots, k \\ &x_{i,} \ge 0, \quad \text{for } i = 1, \ldots, n \end{aligned} \tag{2.12}$$

k the number of constraints.

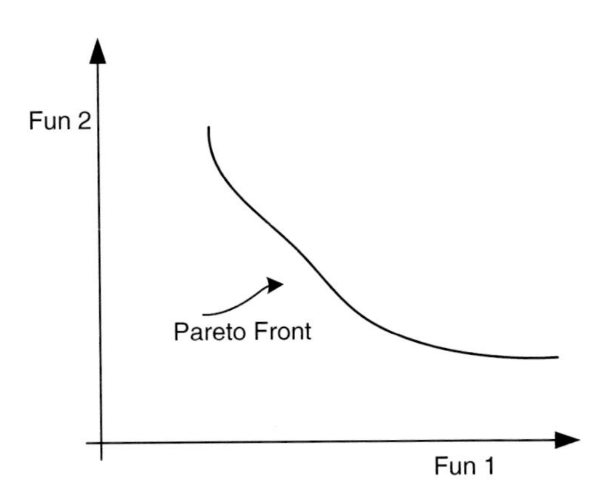

Fig. 2.5 Pareto front

The concept of Pareto dominance defines that a vector x_1 dominates another x_2 for a maximization problem, if for each component of the vector $F_1(x)$ is bigger or equal at each component of $F_2(x)$, but at least one component is greater [14].

$$\begin{aligned} & x_1 \succ x_2 \text{ if,} \\ & f_i(x_1) > f_i(x_2), \quad \text{for at least one } i \text{ of } \exists i \in \{1,\ldots,n\} \\ & \text{and for the remaining comonents of } f_i \\ & f_i(x_1) \geq f_i(x_2), \quad \text{for } i \in \{1,\ldots,n\} \end{aligned} \tag{2.13}$$

A vector solution is called *Pareto optimal* if there is no other vector in the solution space that improve any $f_i(x)$ without deteriorating at least other $f_j(x)$.

From this definition it is constituted the *Pareto set* that represents the objective vectors not dominated with different trade-offs between the objectives.

The Pareto front (PF) or Pareto set (PS) is composed of the set of efficient solutions nondominated. Figure 2.5 represents a MO problem with two OF to minimize, where the solutions to the problem are in feasible region.

$$\begin{aligned} & \mathrm{PS}^* = x_i^* \subseteq \Omega \\ & F\left(x_i^*\right) \subseteq F(\Omega) \\ & \nexists x_m \in \Omega : x_m \succ x_i, \forall x_i \in P^* \end{aligned} \tag{2.14}$$

2.4 Evolutionary Computation

A number of stochastic search strategies are used by researchers to solve the MO problem, the evolutionary algorithms, genetic algorithms, genetic programing, simulated annealing (SA), and ant colony optimization. Those heuristics do not guarantee the identification of optimal Pareto front, but find a good approximation to it.

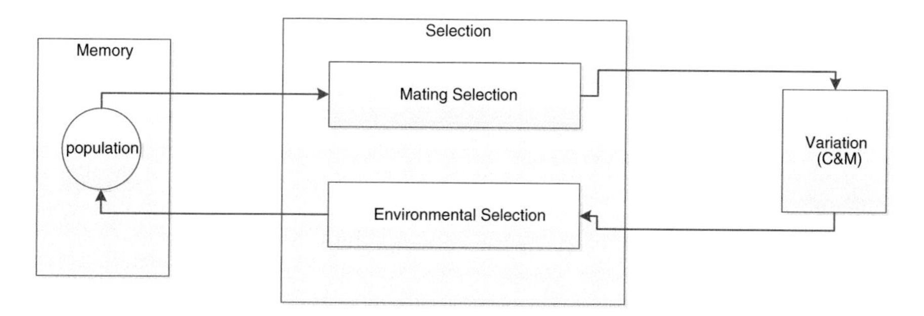

Fig. 2.6 General model of search heuristic

A search heuristic is composed of working memory, selection module, and reproduction model, as represented in Fig. 2.6.

2.4.1 *Memory Module*

This module contains the currently considered population, each time a reproduction is made the new population found is recorded in the memory by replacing the previous one.

2.4.2 *Selection Module*

The selection model are composed of two processes the mating and environmental selection. The mating process picks the most promising individuals from the population and sends these individuals to the reproduction module. After the reproduction process creates new solutions the environment selection determines which of the new solutions are stored in the memory as the new population.

2.4.3 *Reproduction Model*

The variation or reproduction model uses a set of solutions for systematically or randomly genetically changing these solutions to generate a new set of potential better solutions.

2.5 Evolutionary Algorithms

EA are optimization computer algorithms inspired by the biological model of evolution that improve automatically through experience. They use a mechanism that simulates the process of evolution to search in the space of solutions the global optimum of the problem. Using a method to evaluate the quality of randomly generated solutions they progress toward the optimum using a fittest measure to choose the best candidates for reproduction.

The research done in the field proves that it is a heurist capable to solve the optimization problem and find the Pareto set, because EA are capable of processing a set of solutions in parallel and finding a good approximation in a single run [15] (Fig. 2.7).

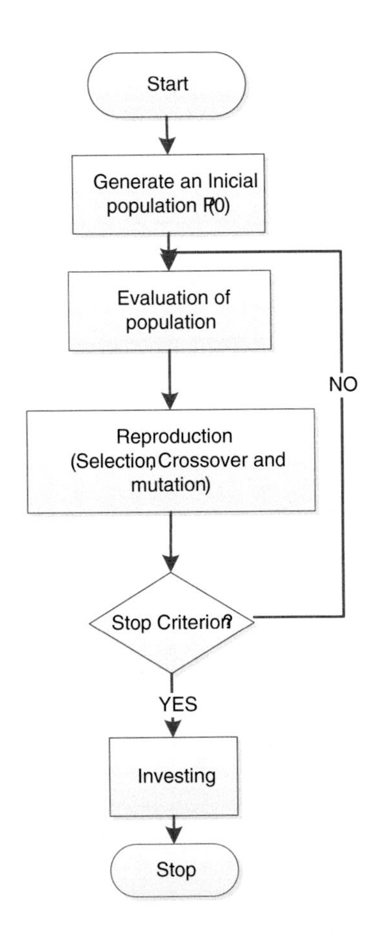

Fig. 2.7 Flow chart of an EA

2.5.1 Representation

EA solutions are represented as a chromosome that is a collection of *genes*, each gene represents one decision variable. The chromosome is an encoding of the solution to the problem, and is called genotype.

2.5.2 Population

The set of solutions that the algorithm finds at each iteration is called population, normally in the first one is used a random population, or a population from a previous simulation to initiate the process can be used.

2.5.3 Quality Indicator

The quality indicator or fitness function is designed to give a quantitative measure of the quality of the chromosome in solving the problem. The selection module uses the evaluation done by the fitness function to select the chromosomes for reproduction. At each iteration it is expected that the average fitness of the population improves meaning that the algorithm is learning.

In single-objective optimization the OF and fitness function are often equal, but for multi-objective different strategies are used to do the fitness, they are aggregation-based, criterion-based, and Pareto-based fitness assignment strategies [14].

2.5.4 Selection

After the fitness is performed the algorithm needs to decide what are the individuals obtained in the reproduction that compose the next population, this is done by a mechanism where the best individuals from the population are selected to be used for reproduction in the next iteration. This process of selection and reproduction steers the search in the direction of the nondominated front at each selection. The common selection operators are the fitness proportionate selection, truncation selection, ranking selection, tournament selection [16].

2.5.5 Variation or Reproduction Operations

Variation operations are realized in the individuals who are selected for reproduction to generate the next population, by the processes of mutation and crossover.

(u) **Crossover**

Recombination or crossover operation uses a certain set of solutions to create a predefined number of new solutions called children, by recombining parts (genes) of each solution [14].

(v) **Mutation**

Mutation is the operation that selects some individuals of the population to reproduction randomly, consists in modifying these individuals by changing some genes according to a given mutation rate.

2.5.6 *Algorithm Design Issues*

According to [14] in MO the goal is to achieve the set of solutions that are as close as possible to the global Pareto optimal front, by minimizing the distance of the solutions generated to the Pareto front. This is obtained by selecting the individuals that are nondominated for them to reproduce. The algorithms need to maximize the diversity of the population obtained to cover the whole Pareto front, this is done by avoiding populations that contain too much identical solutions.

Next, is explained the considerations to take in the design of a multi-objective evolutionary algorithm (MOEA).

(w) **Diversity Preservation**

For a MOEA to preserve the diversity within the current Pareto set approximation, it is needed to incorporate density information in the algorithm to be used in the selection process. This implies that in the selection module the chances of individual's being selected for reproduction decrease, if the density of individuals in its neighborhood is greater. For doing the diversity preservation there are three techniques, the kernel methods, nearest neighbor techniques, and histograms and these techniques are described in [14].

(x) **Elitism**

Elitism addresses the problem of losing good solutions during the optimization process, due to random effects. For solving this problem in the new populations the best elements of the old population are preserved [14].

(y) **Limit Behavior**

Limit behavior of the MOEA is what the algorithms can achieve in terms of performance, when they have unlimited resources, infinite time to continue the evolution and space in memory to store all the solutions.

(z) **Global Convergence**

Global convergence for MOEA is when the Pareto front approximation $PF^{*(t)}$ achieved by the algorithm is practical, identical to the true Pareto front when the number of generations t goes to infinity. In real implementations the algorithms have to deal with limited resources and should guarantee the convergent approximation to the true Pareto front $PF^{*(t)\prime} \subseteq PF^{*}$ [14].

2.5.7 *State-of-the-Art Multi-objective Evolutionary Algorithms*

MOEAs are distinguished from standard EAs by employing the Pareto dominance concept in the fitness evaluation to allow the comparison between individuals based on multiple conflicting objectives.

The first implementation of a MOEA, dated from 1984 with the algorithm VEGA, was a simple genetic algorithm with a modified selection mechanism, developed by Shaffer;from then many new algorithms have been developed. Srinivas and Deb developed NSGA in 1989 that ranked the population with the Pareto dominance and used a dummy fitness value proportional to the population size; MOGA uses a scheme in which each is given a rank proportional to the number of individuals, created in 1993 by Fonseca and Fleming; Horn, Nafpliotis, and Golberg developed NPGA in 1994, this algorithm uses a tournament selection scheme based on Pareto dominance; SPEA uses an external archive for saving the nondominated solutions found in each iteration, which was introduced by Zirzler and Thiele in 1999; PAES developed by Knowles and Corne in 1999 uses an evolutionary strategy with one parent produce on child by mutation; PESA uses hypergrid scheme technique for selection and diversity maintenance, developed in 2000 by Corne, Knowles and Oates. Deb, Pratab, Algarwall, and Meyarivan propose the NSGA-II, whose main characteristic is the computational implementation is easier than NSGA; Erickson, Mayer, and Horn created NPGA-II in 2001, it uses a Pareto approach with tournament selection scheme; SPEA- II introduced by Zitler, Laumanns, and Thiele in 2001 is an improved version of SPEA with an improved fitness assignment; Corne, Jerram, Knowles, and Oates have developed PESA-II in 2001 which uses a selection method where the selective fitness is done to the objective space and not to the individuals; in 2012 Liang, Jane You, Han, Li, proposed the DS-MOEA which uses three alternative operations to perform the learning of the population (evolution in the objective space, evolution in the solution space and self-evolution); and a new algorithm called PSS uses a method to select the nondominated solutions using an global valuation done by fuzzy measures with the user's degree of consideration objective and partial objective evaluation, it was created by Jong Kim, Ji Han,Ye Kim, Seung Choi, Eun Kim in 2012.

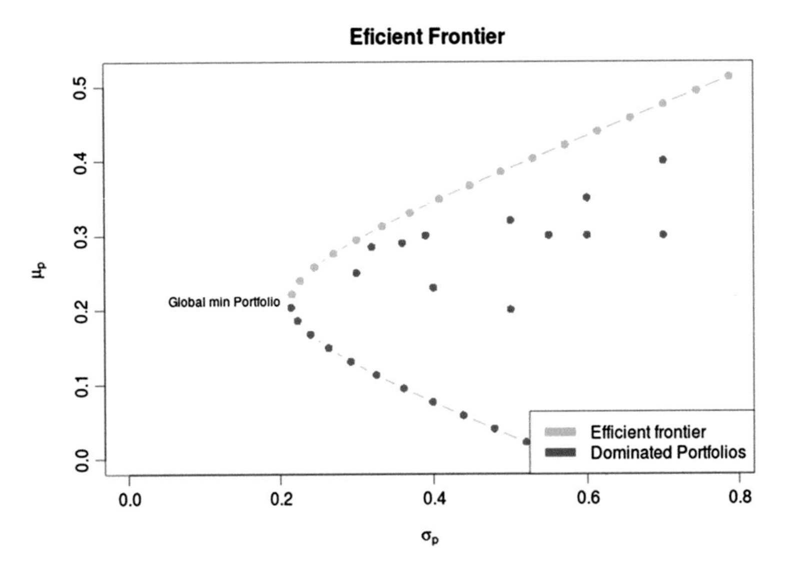

Fig. 2.8 Efficient frontier

2.6 Portfolio Optimization

A portfolio can be defined as a set of selected investments done by its manager. He has at their disposal an amount of capital and a range of assets, to maximize the portfolio return and minimize the risk using asset allocation.[3] The efficient frontier in Fig. 2.8 represents the portfolios with highest expected return for any risk level.

2.6.1 Diversification and Risk of Portfolios

The arbitrage pricing theory using one factor model of Eq. 2.15 is a tool that allows describing and quantifying the factors that affect the rate of return of a security.

$$R = R + \beta F + \varepsilon \tag{2.15}$$

The risk of a stock is the unanticipated part of the return ($\beta F + \varepsilon$) and can be separated in two components, the systematic risk (βF) or market risk due the

[3]Investment strategy that attempts to balance the risk versus return by adjusting the percentage in each asset selected to incorporate the portfolio.

macroeconomic environment that affects the general market, and unsystematic risk (ε) that affects only the company or the industry where it is insert.

This concept applied to the portfolios theory, defines that the return of a portfolio is given by the weight sum of individual returns, Eq. 2.16. It can be noted the diversification effect in the risk reduction of the portfolio in the equation, because the unsystematic risks are independent, the increase of securities in the portfolio decreases the value of the sum of weights of unsystematic risks.

$$R_p = w_1 \times (\bar{R}_1 + \beta F_1 + \varepsilon_1) + \cdots + w_n \times (\bar{R}_n + \beta F_{2n} + \varepsilon_n) \quad (2.16)$$

This theory is applied to the portfolios when it used some diversification to reduce the weight average of the unsystematic risk by diversification.

2.6.2 Mean-Variance Model

The classical mean-variance portfolio optimization model introduced by Markowitz aims to select a set of assets to invest from the space of available assets and determine the fractions w_i of the budget to be invested in each asset selected. In this model the objective is to minimize the risk for each level of expected return of the portfolio.

The Markowitz portfolio optimization can be stated mathematically as follows [17]:

$$\text{Min } \rho(w) = \sum_{i=1}^{n} \sum_{j=1}^{n} \sigma_{ij} w_i w_j \quad (2.17)$$

σ_{ij} Convariance of asset i with the asset j.

The portfolio (P) is a set of real-valued weights (w_i) of the stocks selected from the n available assets in the market.

$$\sum_{i=0}^{n} w_i = 1, 0 \leq w_i \leq 1, \quad i = 1, \ldots, n \quad (2.18)$$

The return of each asset is given by

$$E_i = w_i \times r_i \quad (2.19)$$

r_i valorization of asset i.

The expected return of the portfolio is given by the sum of the expected returns of the individual assets:

$$r_p = \sum_{i=1}^{n} w_i \times r_i \tag{2.20}$$

2.6.3 *The Mean-Variance Cardinality Constrained Portfolio Optimization Model (MVCCPO)*

In the real world, portfolio managers need to deal with a number of realistic constraints that arise from normal business practices and industry regulations [18].

The model MVCCPO is an expanded model of the Markowitz, where it is added two constraints, the cardinality constraint and the quantity constraint.

Cardinality constraint imposes a limit on the number of assets to be selected, it is used by managers to restrict the number of assets in the portfolio to a number that is possible following the economic factors of each asset, and control the transaction costs.

Quantity constraint or buy-in threshold restricts the weight of each asset in the portfolio between lower and upper bounds. The low limit is used to avoid small holdings that do not have almost any effect in the performance of the portfolio, and to prevent to pay higher costs of transaction. The higher limit is used to prevent the excessive exposure of the portfolio to a single asset [19].

Mathematical formulation of MVCCP Model:

$$\text{Min } \rho(x) = \sum_{i=1}^{n} \sum_{j=1}^{n} \sigma_{ij} x_i x_j \tag{2.21}$$

$$\text{Max } r_p(x) = \sum_{i=1}^{n} r_i x_i \tag{2.22}$$

$$\sum_{i=0}^{n} x_i = 1, 0 \le x_i \le 1, \quad i = 1, \ldots, n \tag{2.23}$$

$$\sum_{i=0}^{n} \delta_i \le K \tag{2.24}$$

$$\begin{aligned} l_i \delta_i \le x_i \le u_i \delta_i, &\quad i = 1, \ldots n \\ \delta_i \in \{0, 1\}, &\quad i = 1, \ldots n \end{aligned} \tag{2.25}$$

In this model Eqs. 2.21 and 2.22 are the objective functions to optimize the MOEA, in Eq. 2.23 it is imposed a restriction that the total amount invested need to be equal to the capital available, Eq. 2.24 represents the cardinality constraint, and Eq. 2.25 limits the weight of each investment done ($\delta_i = 1$) to an inferior limit l_i and upper limit u_i.

2.7 State of the Art of Portfolio Optimization

Some real-world problems involve simultaneous optimization of several incommensurable and often competing objectives. Normally, there is not only a single optimal solution, but a set of optimal solutions.

Multi-objective optimization is a method to solve the problem of finding the best solutions when optimizing two or more objectives that are in conflict with each other, subjected to certain constraints [20].

There are many measures of risk and return to evaluate the performance of a portfolio, and these measures can be used as the objectives to be optimized by the EA. The most popular in portfolio management are the compound annual growth rate (CAGR) percentage, managed account report (MAR) ratio, Sharpe ratio, value at risk (VaR), conditional value at risk (CVar), the mean (Portfolio Expected Return), and the variance [13].

Tettamanzi and Loraschi in 1995 describe a MOEA using the Markowitz model, but the measure of risk used is the lower partial moments or downside risk introduced by Harlow [21]. This objective takes into account the downside part of the distribution of returns. The research proves that downside risk make the use of quadratic optimization techniques impossible, because the shape of the OF is non-convex.

Cesarone, Scozzari, and Tardellain (2009) used an algorithm-based approach that starts from a pair of assets in the portfolio and tries to add one each time for the MVCCPO model. The simulations proved that this model of investment has better performance than the Markowitz classical model.

In the work done by Chang, Meade, and Sharaiha in 2000 three heuristics TS, GA, and SA are used to solve the portfolio optimization problem. They also study the problem of finding the efficient frontier using the MVCCPO model. The results prove the existence of cardinality constraint that affects in the shape of the efficient frontier, causing discontinuities in the curve.

In the work Lin and Wang in 2001, NSGA-II is used in the Markowitz's model with constraints of fixed cost and minimum lots. The results show that the efficiency of the GA is undermined without the fitness scaling, and the transaction costs dislocate the Pareto curve in the vertical axis.

Schaerf in 2002 uses MVCCPO model to compare and combine different neighborhood relations in the Pareto front, with local search strategies to find it.

Schyns and Crama in 2003 describe the application of SA for solution of the classic Markowitz model with more realistic constraints, the quantity constraint, cardinality constraint, turnover constraints, and trading constraints. The advantage of using SA over other heuristic methods is the ability to avoid getting trapped in optimal local points, its flexibility and ability to approach global optimality. The important conclusions of this paper are the introduction of trading constraints that are difficult to handle, and there is a trade-off between the quality of the solutions and the time of the simulations to find them.

Lozano and Armañanzas in 2005 uses the heuristics greedy search (GS), SA, and the ACO. For the simulations they used data from five different market indexes. They varied the number of assets (K) in the portfolio for each simulation. The simulations show that fewer assets in a portfolio can represent a higher expected return but it is obtained with a higher variance of the return. They concluded that ACO is a better heurist than the others to obtain portfolios solutions with higher risk more close to the true Pareto front, and the SA fits better for lower risk values.

Clack and Patel in 2007 compared the performance of a standard EA against an Age-Layered Population Structure EA (ALPS EA). They use in the portfolio a basket of 82 stocks of the 100 available. The simulations performed showed that ALPS EA reduces the premature convergence, providing better fittest solutions than the EA.

Ghang, Yang, and Chang in 2009, tested different risk measures in substitution of the mean-variance, one of them is the variance with skewness, developed based on the theory that portfolio return may not be a symmetrical distribution, this means that the distribution of return of individual assets tend to exhibit a higher probability of extreme values, like it has been suggested first by Samuelson in 1958. The results show that MOEO is capable of finding a wider spread of solutions than the others algorithms, and is capable of competing with NSGA-II, SPEA 2, and PAES in portfolio optimization problems.

Hirabayashi, Aranha, Hitoshi in 2009 proposed a GA to generate trading rules based on technical indicators (RSI, MA, percent difference from moving average). The algorithm after entering in a position, will exit it based on the following genes stop loss or take profit optimized. They used this system to trade in the forex market (FX).

Golmakani and Fazel in 2011 used an extended model of Markowitz, with four constraints (minimum transaction lots, sector capitalization lots, cardinality, and quantity constraints). The authors proposed a heuristic called CBIPSO (combination of binary PSO with improved PSO) to solve the portfolio optimization problem of Markowitz.

They compared their heuristic against the GA proposed by Soleimani in 2007. In the simulation they tested different portfolio sizes and expected returns, and they conclude that the CBIPSO outperforms genetic algorithms (GA), which can achieve better solutions in less amount of time.

Hassan and Clack in 2009 [22] tested the combinations of two techniques mating restriction and diversity enhancement in the algorithm SPEA2, to improve the robustness and the diversity of the solutions. To evaluate the quality of the solutions they used the Sharpe ratio.

Casanova in 2010 used a learning classifier system (LCS) in a dynamic learning system to select the stocks to invest based on technical and intuition analysis, the revaluation period RP, average revaluation period (ARP), RSI, MA, DMA are the indicators used for ranking the best stocks for trade, considering the genes parameters (days; minimum value selection of the parameter, variation allowed of the best stock, type of price) for each indicator; with a system for tactical asset allocation called Tradinnova-LCS simulates the intelligent behavior of an investor in a continuous market to form the portfolio. The system tested outperforms all the investment funds analyzed by the INVERCO in the periods of simulation.

Gorgulho, Neves, and Horta in 2011 implemented an expert technical trading system, describing the system architecture and the investment simulator, and used GA to find the solutions. They tested the system against B&H strategy, and random selection, to prove the superiority of the GA system based on technical signals.

The approach of Kaucic presented in 2012 is a trading system based on technical analysis, where an investment module is used to manage a portfolio with long and short positions to generate the so-called long-plus-short portfolio. A technical module is used for detecting overbought/oversold conditions and short-term changes in relative value in contrast to long-term through a learning mechanism using EA that manages the information derived from the technical indicators incorporated.

Pandari, Azar, and Shavazi in 2012 developed a MOEA model with six objectives to optimize, and tested it against the classical model of Markowitz. The conclusion that they arrive is that their model use less risk due to the higher number of objectives optimized by the algorithm.

In Table 2.6 a summary of the different solutions related to the optimization of portfolios using several parameters to describe their main characteristics is presented.

Table 2.6 Overview approaches to portfolio optimization

Reference	Period of simulation	Algorithms utilized	Markets tested	Fitness functions	Constraints	Portfolio analysis	Results obtained
Chang et al. [19]	Mar 1992 to Sep 1997	GA, TS, SA	Hang Seng, DAX, FTSE, S&P, Nikkei	Mean, variance	Minimum lots	Markowitz's model	Best results obtained with the GA Heuristic
Lin and Wang [23]	Mar 1992 to Sep 1997	GA based on NSGA-II and Genocop	Hang Seng index	Mean, variance	Fixed transaction costs minimum lots	Markowitz's model	The proposal GA solves the portfolio selection problem efficiently
Schaerf [24]	NA	TS, SA, LS	Hang Seng, DAX, FTSE, S&P, Nikkei	Average percentage loss	Cardinality, quantity	Markowitz's model	The Tabu search is the heuristic that achieves the best Pareto curve
Schyns and Crama [25]	Jan 6, 1988 to Apr 9, 1997	SA	151 US Stocks	Mean, variance	Floor ceiling, turnover, trading and quantity	Markowitz's model with	The SA is able to handle more classes of constrains than other heuristics
Lozano and Armañanzas [26]	Mar 1992 to Sep 1997	GS, SA, ACO	Hang Seng, DAX, FTSE, S&P, Nikkei	Mean, variance	Cardinality, quantity	Markowitz's model	They obtained a portfolio with a return of 3 and risk 0.1
Clack and Patel [27]	May 31, 1999 to Dec 31, 2005	ALPS system incorporated in GP	FTSE 100	Sharpe ratio		Nonlinear model	The ALPS GP obtained a return of 50 %, and the standard GP a return of 33 %

(continued)

Table 2.6 (continued)

Reference	Period of simulation	Algorithms utilized	Markets tested	Fitness functions	Constraints	Portfolio analysis	Results obtained
Ghang et al. [28]	From Jan 2004 to Dec 2006	GA	Hang SENG, FTSE, S&P	Mean, variance, semivariance, variance with skewness		Markowitz's model	The higher return obtained for S&P was 0.0023 with a risk 0.0008
Chen et al. [29]	Mar 1992 to Sep 1997	MOEO algorithm	Hang Seng, DAX, FTSE, S&P, Nikkei	Mean, variance	Cardinality, quantity	Markowitz's model	Best performance obtained with a MOEO with a return of 0.00859 and risk 0.000417
Anagnostopoulus and Mamanis [30]	Mar 1992 to Sep 1997	NPGA2, NSGA-II, PESA, SPEA2, e-MOEA	Hang Seng, DAX, FTSE, S&P, Nikkei	Mean, variance	Cardinality, quantity	Markowitz's model	The SPEA2 is superior than other MOEA
Casanova [31]	2005 to 2009	LCS model	IBEX 35	ROI		Technical analysis	15.3 %
Gorgulho et al. [32]	Jan 06, 2003 to Jan 06, 2009	GA	DJI	ROI		Technical trading system	60 %
Kaucic [33]	Jan 25, 2006 to Jul 19, 2011	EA	DJI	Information ratio, omega, Sortino ratio		Technical trading system	70 %, 125 %, 119 %
Pandari et al. [34]	Mar 2002 to Mar 2008	MOEA	Tehran Stock	Cumulative return, mean return		GA model using Sharpe ratio, Markowitz's model	600 %, 350 %

2.8 Conclusion

In this chapter major investment approaches were presented, in particular the value investing, where it is explained their components, such as financial analysis performed using financial statements of the companies, the economic indicators that influence the business sector, and macroeconomic indicators to take into account in forecasting the economic cycle, for investors have a better perspective of the future and profitability of investments.

It has explained the method of investment of Warren Buffet, and the DCF methods used to evaluate companies depending on the cash flow generated and growth rates. It was presented the different types of markets, and the necessary components to build a complete trading system.

The optimizing concepts of multiple objectives, such as Pareto curve, and the Pareto set, were discussed in this work in a way to be applied in the development of the MOEA. It was made a brief introduction to evolutionary algorithms and how to implement them. It is explained the most important considerations to have in implementation and possible problems of the EA.

After a description of the state of the art of multi-objective EA, are introduced the mean-variance model, and the mean-variance cardinality constrained portfolio optimization model (MVCCPO). It is presented the state of art of the investigation than with portfolio optimizations, the models of investment, approaches to manage the portfolios used, and the objectives to optimize.

The active strategies of investment based mainly on technical analysis, demonstrate that there exist a number of possible applications of intelligent computing applied to investments in stocks, and can achieve good investment returns.

References

1. B. Graham, D.L. Dodd, *Security Analysis* (Mcgraw Hill, New York, 1934)
2. D. Hillier, S. Ross, W. Randolph, J. Jeffrey, B. Jordan, *Corporate Finance*, First European Edition (Mcgraw-Hill, New York, 2010)
3. M. Buffett, D. Clark, *Warren Buffett and the Interpretation of Financial Statements* (Simon & Schuster, New York, 2008)
4. R.S. Pindyck, D.L. Rubinfeld, *Microeconomics* (PERSON Prentice Hall, Upper Saddle River, 2006)
5. T.R. Robinson, H.V. Greuning, E. Henry, M.A. Broihahn, *International Financial Statement Analysis* (Wiley, Hoboken, 2009)
6. Z. Bodie, A. Kane, A. Marcus, *Investments* (McGraw-Hill, New York, 2010)
7. O. Blanchard, *Macroeconomics* (PERSON Prentice Hall, Upper Saddle River, 2006)
8. P. Lynch, J. Rothchild, *One Up On Wall Street* (Simon & Schuster, New York, 1989)
9. R. Hagtrom, *The Warren Buffett Way* (Wiley, India, 1994)
10. S. Achelis, *Technical Analysis from A-to-Z* (Vision Book, New Delhi, 2000)
11. V.K. Tharp, *Trade to Financial Freedom* (Mcgraw-Hill, New York, 2007)
12. C. Faith, *Way of the Turtle* (McGraw-Hill, New York, 2007)

13. K. Metaxiots, K. Liagkouras, Multiobjectve evolutionary algrithms for portfolio management: a comprehensive literature review. Expert Syst. Appl. **39**, 11685–14085 (2012)
14. E. Zitzler, M. Laumanns, S. Bleuler, A tutorial on evolutionary multiobjective. Metaheuristics Multiobjective Optim. **535**, 3–37 (2004)
15. E. Zitzler, L. Thiele, An evolutionary algorithm for multiobjective optimization: the strength pareto approach (Computer Engineering and Networks Laboratory (TIK), Swiss Federal Institute of Technology Zürich (ETH), 1998)
16. G.N.A. Hassan, *Multiobjective genetic programming for financial portfolio management in dynamic environments* (Doctoral dissertation, UCL (University College London)) (2010)
17. H. Markowitz, Portfolio selection. J. Finan. **7**, 77–91 (1952)
18. P. Skolpsdungket, K. Dahal, N. Harnpornchai, Portfolio optimization using multi-objective genetic-algorithms. IEEE Congr. Evol. Comput. 516–523 (2007)
19. T. Chang, N. Meade, Y. Sharaiha, Heuristics for cardinality constrained portfolio optimisation. Comput. Oper. Res. **27**, 1271–1301 (2000)
20. C.M. Fonseca, P.J. Fleming, Genetic algorithms for multiobjective optimization: formulation, discussion and generalization. In *Proceedings of the Fifth International Conference on Genetic Algorithms* (1993), pp. 416–423
21. H. Harlow, Asset allocation in a downside-risk. Finan. Anal. J. **47**, 28–40 (1991)
22. G. Hassan, C.D. Clack, Robustness of multiple objective GP stock-picking in unstable financial markets: real-world applications track. In *Proceedings of the 11th Annual conference on Genetic and evolutionary computation* (2009), pp. 1513–1520. ACM
23. D. Lin, S. Wang, A multiobjective genetic algorithm for portfolio selection problem. in *Proceedings of ICOTA* (2001)
24. A. Schaerf, Local search techniques for constrained portfolio selection problems. Comput. Econ. **20**, 177–200 (2002)
25. M. Schyns, Y. Crama, Simulated annealing for complex portfolio selection problems. Eur. J. Oper. Res. **150**, 546–571 (2003)
26. J. Lozano, R. Armañanzas, A multiobjective approach to the portfolio optimization problem. IEEE Congr. Evol. Comput. **2**, 1388–1395 (2005)
27. C. Clack, S. Patel, ALPS evaluation in financial portfolio optimization. IEEE Congr. Evol. Comput. 813–819 (2007)
28. T.J. Ghang, S.C. Yang, K.J. Chang, Portfolio optimization problems in different risk meusures using genetic algorithm. Expert Syst. Appl. **36**, 10529–10537 (2009)
29. M.R. Chen, J. Weng, X. Li, Multiobjective extremal optimization for portfolio optimization problem. Intelligent Computing and Intelligent Systems, 2009. ICIS 2009. IEEE International Conference on, vol. 1. IEEE 552–556 (2009)
30. K. Anagnostopoulos, G. Mamanis, The mean-varaince cardinality constrained portfolio optimization problem: an experimental evaluation of five multiobjectives evolutionary algorithms. Expert Syst. Appl. **38**, 14208–14217 (2011)
31. I.J. Casanova, Trading-LCS: dynamic stock portfolio decision-making assistance model with based machine learning, IEEE Congr. Evol. Comput. 1–8 (2010)
32. A. Gorgulho, R. Neves, N. Horta, Appying a GA kernel on optimizing technical analysis rules for stock picking and portfolio composition. Expert Syst. Appl. **38**, 14072–14085 (2011)
33. M. Kaucic, Portfolio management using artificial trading systems based on technical analysis. Genet. Algorithms Appl. Chapter 15, (INTECH Open Access Publisher, 2012)
34. A.R. Pandari, A. Azar, A.R. Shavazi, Genetic algorithms for portfolio selection problems with non-linear objectives. Afr. J. Bus. Manage. **6**, 6209–6216 (2012)

Chapter 3
System Architecture

Abstract In this chapter, the components necessary to evaluate companies, like the financial analysis and the meaning of the important item in each statement, will be explained. A brief explanation about technical analysis and its use to define buy and sell positions will also be presented.

The system developed uses the value investing approach with a technical trading system to invest in the stock market. For the system to work it is necessary to obtain the financial market data to perform the training and real test simulations.

Due to the wide availability of information and data in the US markets, it is essential to focus on the really important information. The financial information provided, as well as the economic analysis of the sector, and the management of the company, are the most relevant data information to the analysts. In this work, it is used the quarterly financial information available, and the adjusted closing prices of the companies of S&P 500.

This chapter presents the architecture of the system, first in the Sect. 3.1 the algorithm architecture is explained, Sect. 3.2 describes how the financial statements and quotations of listed companies in the S&P 500 are obtained, in Sect. 3.3 the financial statements and how to interpret them is explained, then Sect. 3.4 presents the financial ratios used in the chromosomes to evaluate the companies, and the Sect. 3.5 describes the technical analysis used.

3.1 Algorithm Architecture

The system architecture is presented in Fig. 3.1 and is constituted by three main modules, the investor simulator, the optimizer, and the data. The main blocks of the architecture are the data block that is accessed by the investment simulator to test the strategies, and the optimization block to implement the MOEA. The

A.D. Silva et al., *Portfolio Optimization Using Fundamental Indicators Based on Multi-Objective EA*, SpringerBriefs in Computational Intelligence,
DOI 10.1007/978-3-319-29392-9_3

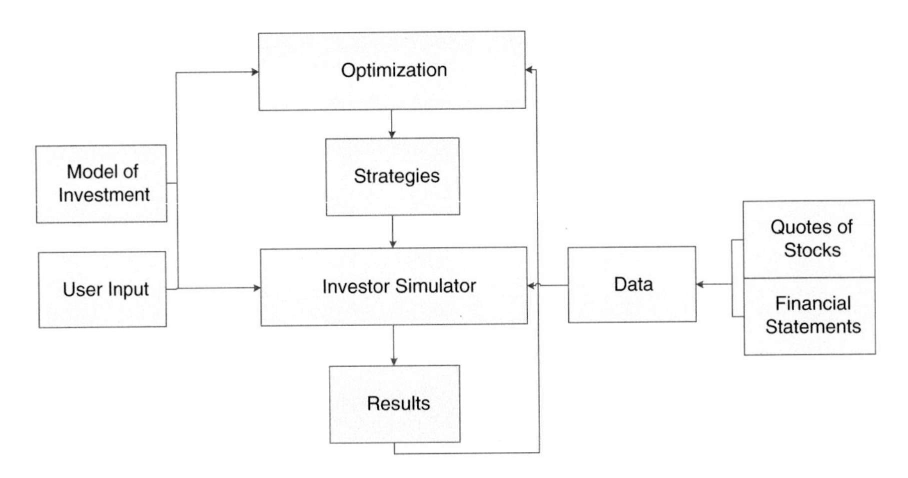

Fig. 3.1 System architecture

optimization block uses the results return of investment (ROI) and variance, obtained in the investor block to calculate the fitness function and evaluate the strategies. It also selects them for reproduction and applies the methods of crossover and mutation to create news chromosomes.

The data block uses the financial statements and stock quotes to calculate the ratios used by the investor simulator. The investor simulator block tests the strategies obtained in the optimization block, depending on the investment model and using the inputs given by the user, for getting the data for testing. It evaluates in each day the stocks, and calculates the ROI of the portfolio from the beginning of the simulation. With the recorded values for each day of the ROI, it is then calculated the monthly variance of the return.

Each strategy is tested in the simulator using the retrieved time series, with a window that slides 1 day at each increment. The selected financial ratios are evaluated, and the computations are performed for doing the trading decisions of the day. After finishing the time frame of each period of training, it is recorded the results, and performed the population reproduction. This process is repeated to achieve the number of iteration that signals the end of the training. After the conclusion of the training of the population the external file with the best solutions is used in the real test.

The system uses the following parameters, the period of training, and the period of real test, also transactions costs are included at 2 % of the stock value.

3.1.1 Architecture of the Evolutionary Algorithms

In Fig. 3.2 a flowchart with the architecture of the algorithm is presented, where the system required some specification to work, the period of training, real test period,

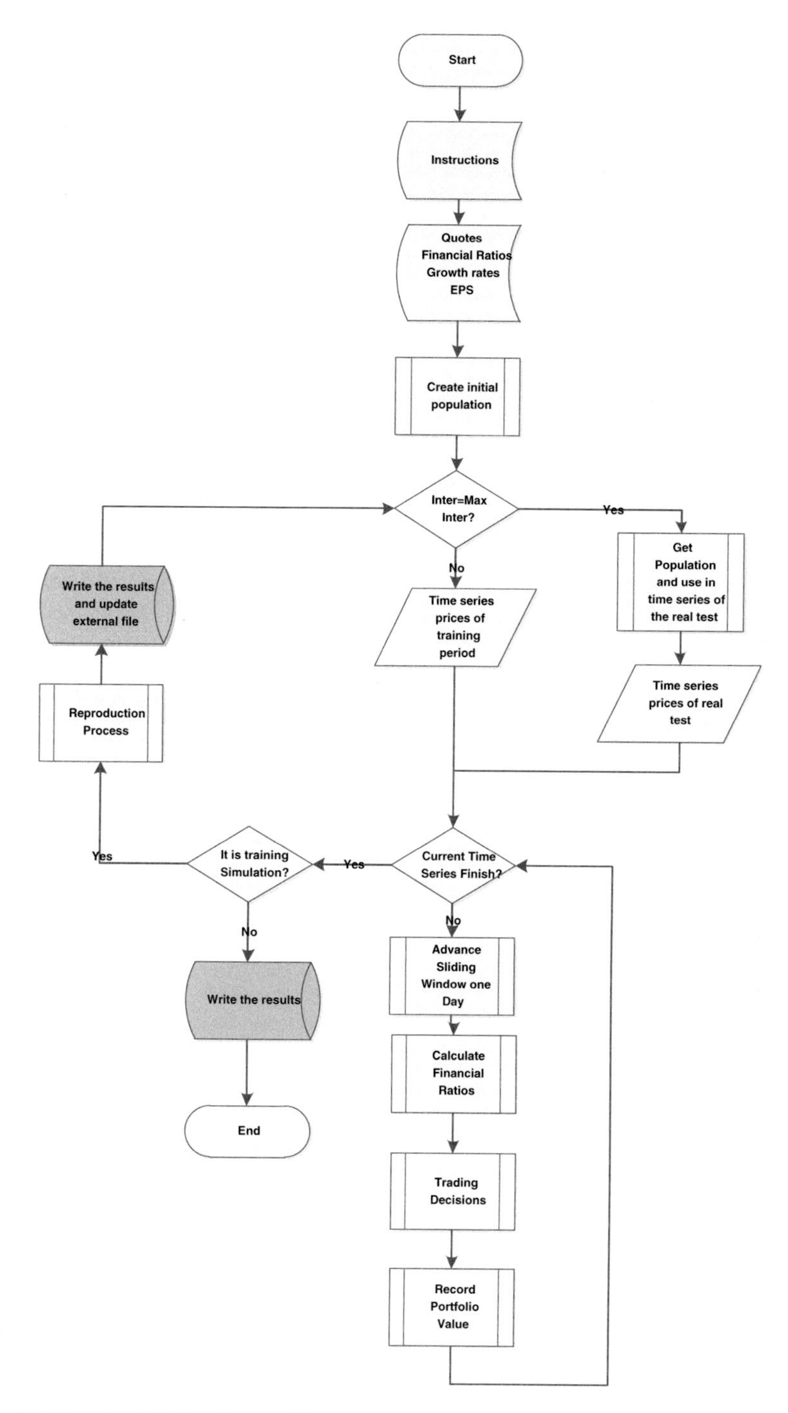

Fig. 3.2 Flowchart of algorithm architecture

commission percentage to charge in each transaction performed, and the number of iterations to perform the training of the population. Next is accessed the data to the algorithm to work in the periods defined, before the start of training is created a random population. For test the chromosomes is created time series with all the stocks prices, where are executed the tasks to manage the portfolio, as calculate the financial ratios, and trading decisions each time the window advance on day. In the end each iteration is recorded the results (return and variance) of each chromosomes for the reproduction and selection process. When number of the iterations arrive to the maximum number the training is finish, and the algorithm perform a real simulation of the population in the external file (the best chromosomes obtained in the training) for the period defined in the specifications. After tested all chromosomes are recorded the results obtained.

3.2 Obtaining the Financial Data

To perform the necessary studies in this work, it was necessary to obtain the financial statements of the companies and their quotations. It was chosen the Yahoo finance database to extract the necessary data, because it allows the download of the Excel files with the market information of each company and have on their site quarterly and annual financial statements for each company. To get the data, a program was developed in C++ in eclipse in the Linux environment to download the quotes of the selected companies, and the respective financial statements. The flowcharts in Fig. 3.3 describe the work of the program mentioned before.

3.3 Financial Statement Analysis

In Chap. 2, some techniques for valuation of stocks were presented, but the input variables of the models are estimates of futures dividends and earnings, for the investor this is based on economic projections that can be very uncertain, the only correct information available is the financial accounting, to estimate the intrinsic value of stocks.

3.3.1 The Income Statement

The income statement is a demonstration of the profitability of the company, over a period of time. The revenues obtained from the operations, the expenses incurred, and the net earnings are reported.

The expenses are separated into four parts, the first is the cost of goods sold, these are the direct cost of generating the products, second are general and

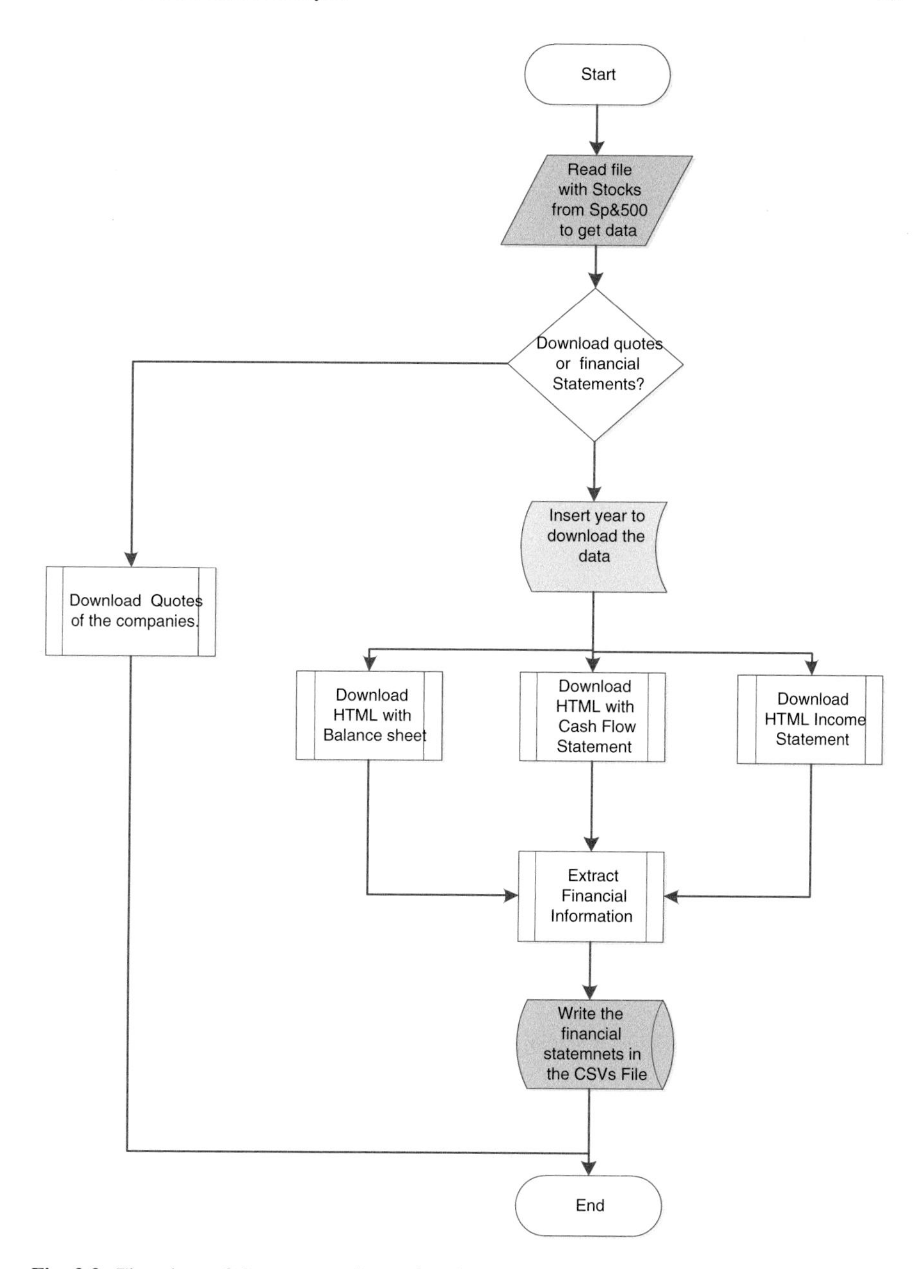

Fig. 3.3 Flowchart of the program for getting the data

administrative costs that are indirect costs related to the production, the third is the interest expenses of thc debt, recorded as financial cost, and the final is the taxes charged by the government.

In Millions of USD (except for per share items)	12 months ending 2013-06-30
Revenue	77,849.00
Other Revenue, Total	-
Total Revenue	77,849.00
Cost of Revenue, Total	20,249.00
Gross Profit	57,600.00
Selling/General/Admin. Expenses, Total	20,425.00
Research & Development	10,411.00
Depreciation/Amortization	-
Interest Expense(Income) - Net Operating	-
Unusual Expense (Income)	208.00
Other Operating Expenses, Total	-
Total Operating Expense	51,293.00
Operating Income	26,556.00
Interest Income(Expense), Net Non-Operating	-
Gain (Loss) on Sale of Assets	-
Other, Net	194.00
Income Before Tax	27,052.00
Income After Tax	21,863.00
Minority Interest	-
Equity In Affiliates	-
Net Income Before Extra. Items	21,863.00
Accounting Change	-
Discontinued Operations	-
Extraordinary Item	-
Net Income	21,863.00

Fig. 3.4 Annual income statement of Microsoft

Figure 3.4 shows the income statement of Microsoft, where in the top are the revenues from operations, next are the costs, by subtracting the costs to revenue is obtained the *operating income or loss,* or *EBIT.*[1] To obtain the *Net Income* in the last row it is extract the interest expenses and taxes paid to the *EBIT.*

3.3.2 Balance Sheet Statement

The balance sheets give the financial condition of the company at particular moment, where it lists all the assets and debts at the time of the report occurred. In the first part of the report are listed the assets from the more liquid to the less, they are divided into current assets, and long-term assets. Next are the part of liabilities that are arranged similar to the short-term debts which came first, and the long-term debts after.

The difference between total assets and liabilities is the stockholders' equity, these are the book value of the company. In the last line of the balance sheet are the

[1]This is a measure of profitability of the firms operations.

In Millions of USD (except for per share items)	As of 2013-06-30
Cash & Equivalents	3,804.00
Short Term Investments	72,971.00
Cash and Short Term Investments	76,775.00
Accounts Receivable - Trade, Net	17,486.00
Receivables - Other	-
Total Receivables, Net	17,486.00
Total Inventory	1,938.00
Prepaid Expenses	-
Other Current Assets, Total	5,267.00
Total Current Assets	101,466.00
Property/Plant/Equipment, Total - Gross	22,504.00
Accumulated Depreciation, Total	-12,513.00
Goodwill, Net	14,655.00
Intangibles, Net	3,083.00
Long Term Investments	10,844.00
Other Long Term Assets, Total	2,392.00
Total Assets	142,431.00
Accounts Payable	4,828.00
Accrued Expenses	4,117.00
Notes Payable/Short Term Debt	0.00
Current Port. of LT Debt/Capital Leases	2,999.00
Other Current liabilities, Total	25,473.00
Total Current Liabilities	37,417.00
Long Term Debt	12,601.00
Capital Lease Obligations	-
Total Long Term Debt	12,601.00
Total Debt	15,600.00
Deferred Income Tax	1,709.00
Minority Interest	-
Other Liabilities, Total	11,760.00
Total Liabilities	63,487.00
Redeemable Preferred Stock, Total	-
Preferred Stock - Non Redeemable, Net	-
Common Stock, Total	67,306.00
Additional Paid-In Capital	-
Retained Earnings (Accumulated Deficit)	9,895.00
Treasury Stock - Common	-
Other Equity, Total	-117.00
Total Equity	78,944.00
Total Liabilities & Shareholders' Equity	142,431.00
Shares Outs - Common Stock Primary Issue	-
Total Common Shares Outstanding	8,328.00

Fig. 3.5 Annual balance sheet statement of Microsoft

total common shares outstanding, representing the number of shares in the market of the company. In Fig. 3.5 is the example of annual balance sheet statement, where it can be seen the assets and liabilities hold by the microsoft.

3.3.3 Cash Flow Statement

The cash flow statement follows the cash transactions and records them. This provides important evidence of the wealth of the company, by revealing the sources

In Millions of USD (except for per share items)	12 months ending 2013-06-30
Net Income/Starting Line	21,863.00
Depreciation/Depletion	3,755.00
Amortization	-
Deferred Taxes	-19.00
Non-Cash Items	4,609.00
Changes in Working Capital	-1,375.00
Cash from Operating Activities	28,833.00
Capital Expenditures	-4,257.00
Other Investing Cash Flow Items, Total	-19,554.00
Cash from Investing Activities	-23,811.00
Financing Cash Flow Items	199.00
Total Cash Dividends Paid	-7,455.00
Issuance (Retirement) of Stock, Net	-4,429.00
Issuance (Retirement) of Debt, Net	3,537.00
Cash from Financing Activities	-8,148.00
Foreign Exchange Effects	-8.00
Net Change in Cash	-3,134.00
Cash Interest Paid, Supplemental	371.00
Cash Taxes Paid, Supplemental	3,900.00

Fig. 3.6 Annual cash flow statement of Microsoft

of cash flow. An illustrative example is presented in Fig. 3.6. The first line represents the net income from the report of income statement for the same period.

The cash flow statement separates cash from operations, financial, and investing activities, this allow an investor to see if a good report of income statement, is due to the operations of the company or to a single event that will not repeat, like cash flow from a sell of fixed assent. The section of financial activities list the cash flow realized or spends with issue or acquired debt and equity.

3.4 Financial Ratios

From the financial statements information it is possible to calculate many financial ratios, which are quantitative measures that allow the analysis of a company in terms of profitability, liquidity, debt, and growth. They are used to compare companies inside the same industry and to draw conclusions about the best companies to invest. Then, it will be presented the ratios used in this work as well as its desired function for select a company.

3.4.1 Debt Ratio

An investor can find what appear to be good company to invest because the income statements presents good results, including *net income* increasing with a high *Profit*

Table 3.1 Debt—ratios

Debt ratio			
Stock	2011	2012	2013
JNJ	0.47	0.50	0.45
KO	0.62	0.60	0.57
MSFT	0.45	0.45	0.47
AAPL	0.40	0.33	0.34
ADI	0.26	0.26	0.28
ADP	0.81	0.80	0.82
AMZN	0.63	0.69	0.74

Margin, and the company is traded in the market with a low PER ratio. This draws the conclusion that it is a profitable company that is undervalued. But this might not be true, there might be a problem with leverage that makes a company appear more profitable than in reality it is, and that it has a competitive advantage.

To filter this type of company it is used the *Debt Ratio*, companies that have high debt compared to its assets are at greater risk of going bankrupt in the occurrence of an adverse economic cycle, or suffer a reduction in their profits by increasing interest rates of their debts. Usually companies with low *Debt Ratio* are in extremely competitive industries with constant need for innovation and updates of its products and production processes, which are carried out using external financing.

$$\text{Debt ratio} = \frac{\text{Total Debt}}{\text{Total Assets}} \tag{3.1}$$

By analyzing the *Debt-ratios* of some companies in Table 3.1 can be verified differences in the capital structures from one company to another. The companies valorization with different debt ratios in Fig. 3.7, demonstrates that it is needed to take in consideration the capital structure to predict the companies with higher potential of return.

3.4.2 *Return on Equity*

Return on Equity is an accounting measure of the performance of the company in generating profits. This ratio is used by investors to select companies that maximize the investment made in them; this means for every dollar invested they are capable of creating a net profit greater in terms of percentage of capital invested. If managers retained the earnings they will be reinvested in the business with a higher return rate, than if distributed as dividends to the shareholders. The ROE is obtained by three factors, the operational efficiency, the efficient use of assets, and the financial leverage, this can be seen in Eq. (3.2). The objective is to choose

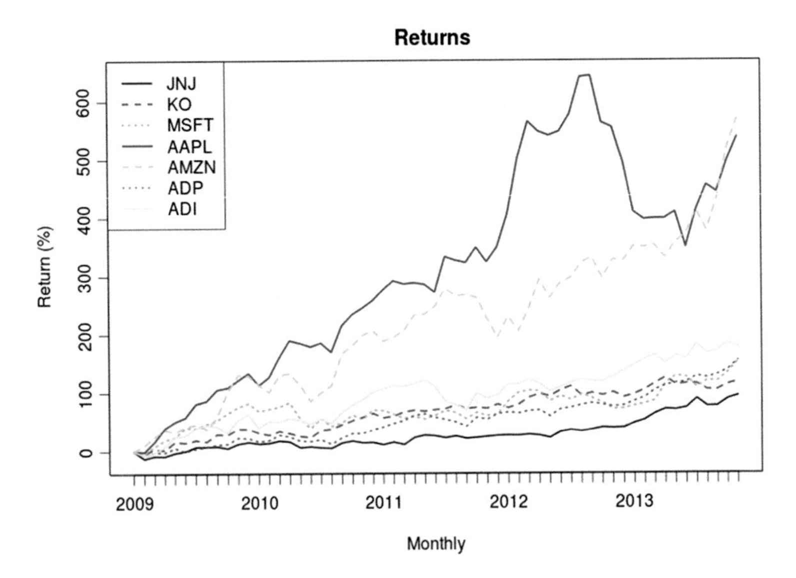

Fig. 3.7 Compound returns of companies of S&P500

companies with higher ROEs, where this value is obtained by the *profit margin* and *asset turnover*, and not obtained through excessive leverage [1]. This is the type of companies that have a competitive advantage, normally monopolistic in their sector of activities, and are traded with high per share earnings.

$$\text{ROE} = \frac{\text{Net Income}}{\text{Total Equity}} = \text{Profit margin} \times \text{Total asset turnover} \times \text{Equity multiplier} \tag{3.2}$$

In Table 3.2 are the values of ROE of some companies from different sectors, can be noticed that companies that belong to the same industry as is the case of Apple and Microsoft have similar ROE.

Table 3.2 Average 5 years of ROE

Stock	Average's 5 year ROE
JNJ	23.2
KO	31.1
MSFT	36.9
AAPL	36.3
ADI	18.2
ADP	23.1
AMZN	16.6

3.4.3 *Profit Margin*

This ratio measures the profitability of the company, calculating as the percentage of the revenue retained after paying the operating, administrative, financial costs, and taxes.

$$\mathrm{PM} = \frac{\text{Net Income}}{\text{Revenue}} \tag{3.3}$$

When analyzing the evolution of the *Revenue* with the *Profit Margin* it is possible to make valid predictions about the future of the company. As observed in Fig. 3.8, the growth trend of the PM along with the growth of Apple's revenue from 2009 to 2012 is reflected in the stock market with a valorization over 400 % for the period in question, having been one of the biggest valorizations in Index S&P 500. This is a method to find the companies with the best businesses that have durable competitive advantage, allowing the increase of the profit margin at the same time with the increase of revenue [2].

3.4.4 *Price Earnings Ratio*

Price earnings ratio established a relationship between the share price and the company's profits are used to measure company levels of overvalued or undervalued. PER tends to be lower for companies, such as slow growers and higher for fast growers because in the share price is incorporated investor expectations regarding to the future. The best use of the ratio is to find the cheapest company in a sector, but

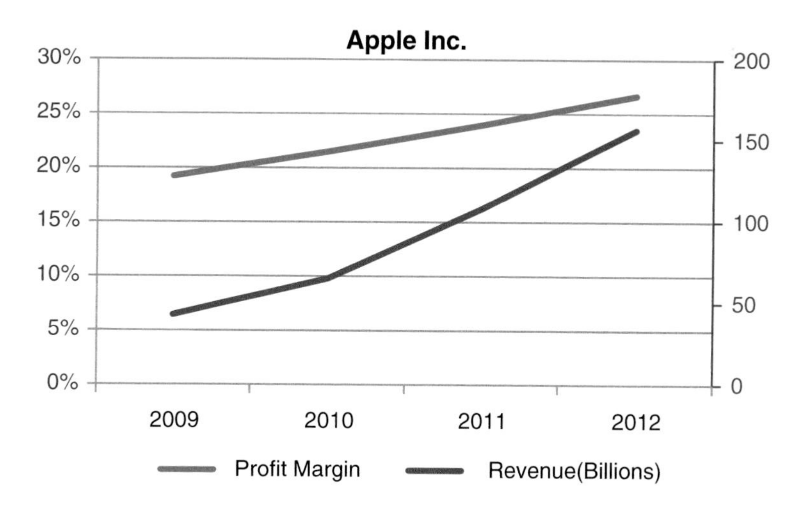

Fig. 3.8 Comparing profit margin and revenue of Apple Inc

can be used to compare companies in different business sectors. It is possible to compare the historical record of PER in order to get a better perspective on the levels of overvaluation and undervaluation of each stock, avoid higher PER ratios is usually good investment policy, as it normally signal overvalued companies. For a company with a PER of 30 or greater it is necessary in the coming years to achieve high rates of profit growth to be a profitable investment. This means that an investor is available to pay a higher price per earnings if expects higher earnings growth.

(a) Price/earnings to growth (PEG-ratio)
The PEG-ratio is a variation of PER that take into account the growth of earnings in the valuation of the company, meaning that a PEG below one signifies undervaluation. It is a good policy to calculate the two ratios to have a better picture about the company, in terms of undervalue of the respective earnings, and in terms of growth.

$$\text{PEG} = \frac{\text{PER}}{\text{Annual EPS Growth}} \tag{3.4}$$

(b) Relation between P/E ratio and ROE
Using the dividend discount model of Eq. (2.5) it can be proved that the P/E ratio increases with the ROE. Considering the dividend is equal to earnings less a rate of reinvested earnings R_i in the company, the dividend is given by Eq. (3.5).

$$D_1 = \text{EPS}_1(1 - R_i) \tag{3.5}$$

The growth rate g is given by the rate of reinvestment of the earnings times the return of equity of the company (ROE).

$$g = R_i \times \text{ROE} \tag{3.6}$$

Substituting Eqs. (3.5) and (3.6) in the Eq. (2.5) to get the Eq. (3.7) that gives the intrinsic price of the share for a company with constant growth rate. Equation (3.8) relates the PER to the ROE, by demonstrating that companies with goods ROE have higher PER.
Companies with higher reinvestment rates will have higher growth rates and this is reflected in the PER like said previously. It is common that high R_i, happened when the ROE of the company exceeds the discount rate, meaning that it is better for the shareholder that the company reinvest its earnings in the business, or in a new project, or buyback shares, because the money received as dividends when reinvested at the better rate of return that the investor has is lower than return of capital given by the company.

$$P_0 = \frac{\text{EPS}_1(1 - R_i)}{R - \text{ROE} \times R_i} \tag{3.7}$$

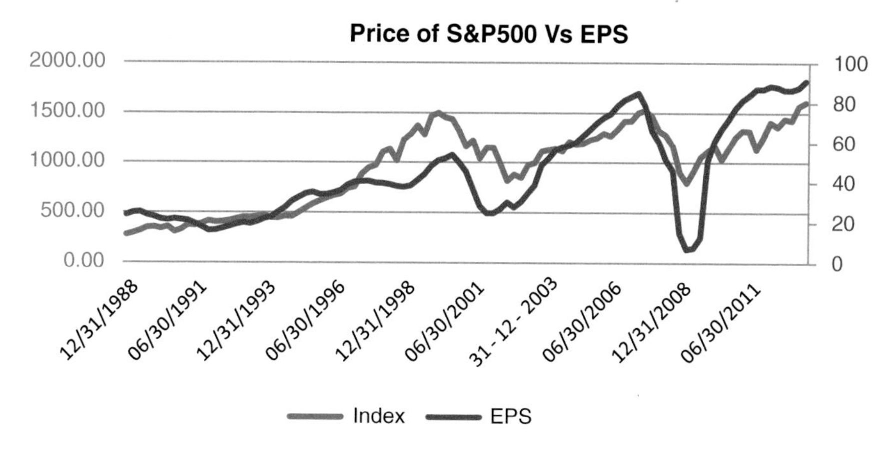

Fig. 3.9 EPS of the market S&P 500 versus index price

$$\mathrm{PER} = \frac{P_0}{\mathrm{EPS}_1} = \frac{(1 - R_i)}{R - \mathrm{ROE} \times R_i} \tag{3.8}$$

(c) The P/E market

The PER of the stock market is calculated by the aggregation of all PERs of the companies, it is indicator used to determining whether the market is overvalued or not. Interest rates influence the overall value of the market, and it is PER, low interest rates increase the real returns of future years, thereby contributing to an increase in investors' expectations, and causing a valuation of the general market. In Fig. 3.9 is represented the month values of EPS and the price of the index S&P 500 from 1988 to 2013, it can be seen that in the years of 2001 and 2007 when there was a major difference between the two trends, the market started a correction. Meaning that when the market is overvalued in relation to its earnings, a bear market occurred, or a higher growth of the economy needs to happen to maintain the higher prices.

3.4.5 *Revenue Growth Rate (RG)*

The percentage of revenue growth is an economic indicator that shows the evolution of the business, two factors are responsible for its increase, the company is a better competitor and is gaining market share to the other competitors, or the company is inserted in a fast growing sector and is growing with it. Although this indicator is important to analyze the company from the growth point of view, it does not give any indication of the profitability of the company. There may be a case where a company has a strong growth in revenues, but due to other factors the net earnings show a decrease, as is the case in recent years of Amazon.

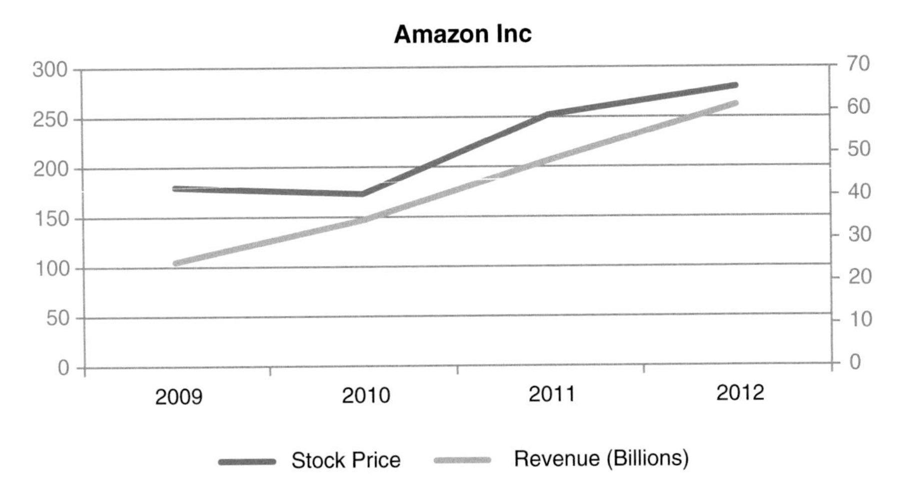

Fig. 3.10 Revenues versus stock price of Amazon

$$RG = 100 * \frac{\text{Revenue}_{\text{Actual}} - \text{Revenue}_{\text{Last Year}}}{\text{Revenue}_{\text{LastYear}}} \tag{3.9}$$

In Fig. 3.10 can be seen the case of Amazon with revenue increases and share price, when other fundamental as the net income did not improve.

3.4.6 Rate of Change in Common Stock Outstanding

Common stock outstanding represents the fundamental ownership hold of the corporation by the shareholders, considering also shares held by the company managers. When the company issues shares this number is added to the previous value in the *Balance sheet*, representing an increase of total number of shares and a distributing of the company's value by a greater number of shares. The monetary value obtained with the selling of these new shares is registered as cash inflow in section of *Issuance (Retirement) of Stock, Net* in the *Cash Flow Statement*. In the case of company repurchases its shares the amount spent is recorded as cash outflow in the same section, this actions removes from circulation shares, and this reduction will be recorded in the *common stock outstanding*.

The reduction of the number of outstanding shares represents an increase in EPS and a decrease of PER ratio. The shares buyback can be a sign that managers, who are better informed than the other participants about the company's business and its future, are considering that the company is undervalued and are optimistic about the future of the company.

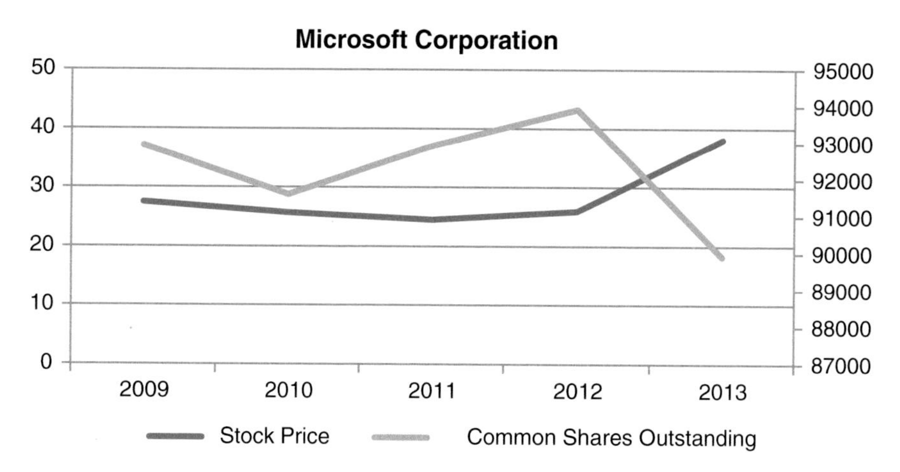

Fig. 3.11 Common shares outstanding of Microsoft

The objective when using this indicator is to find companies that in recent years has repurchased its shares, thus reduced the number of shares outstanding without its stock price suffering any major change as demonstrated by Fig. 3.11.

$$\Delta \text{Common}_{\text{stocks}} = \frac{\text{common stock out}_{\text{Actual}} - \text{common stock out}_{\text{Last year}}}{\text{common stock out}_{\text{Last year}}}. \quad (3.10)$$

3.4.7 *Net Income Growth Rate*

A positive trend of net income shows that the company has consistency and durability in the profits, and a good result obtained in 1year, are not just a result of favorable economic growth of the GDP or a financial engineering. When comparing the trends of net income and stock price, divergences can be found between both, and can be an indication of undervaluation of the company, if the net income trend shows a positive tendency and the stock prices do not show a similar behavior. As observed in Fig. 3.12 the Wal-Mart net income growth in the years from 2007 to 2012, but the stock price between 2008 and 2010 decline from their top.

$$\Delta\,\text{Net Income} = 100 * \frac{\text{Net Inc}_{\text{Actual}} - \text{Net Inc}_{\text{Last Year}}}{\text{Net Inc}_{\text{Last Year}}}. \quad (3.11)$$

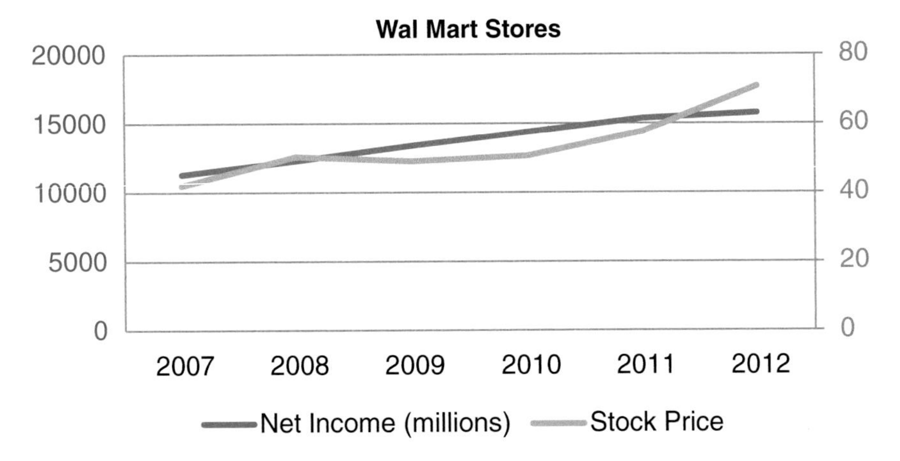

Fig. 3.12 Stock price and net income of Wal-Mart stores

3.5 Technical Analysis

The function of technical analysis is predicting patterns in stocks prices and exploring it. The technical analysts believe that prices are gradually close of its intrinsic value and they use volume data as well as price history to predict the market. One of the principal applications of this analysis is the use of moving averages to find trends in the market prices.

3.5.1 Moving Average

There exist a great number of different types of moving averages, depending of the calculated method, but it functions and interpretations are the same, in this work only the simple moving average (SMA) are discussed and used with the evolutionary algorithms.

Moving average is a technical indicator composed by series of numbers that represent average value of a security′s price, over a set period. The values are the means from several consecutive subsets from the set selected by the period size of the MA, as defined by Eq. (3.12). To calculate the mean values can be used the close, high, low, open and Adj.Close prices. The Adj.Close prices are used by the algorithms designed for calculating the SMA for the stocks of the Index.

$$\mathrm{SMA}_t = \frac{P_t + P_{t-1} + \cdots + P_{t-n}}{n} \tag{3.12}$$

Fig. 3.13 Trend of SP&500 with two SMA of 200 and 150 days

The MA function is used to eliminate noise in the stock price, thus allowing the analyst to better determine the trend. MA is also used in trading systems as trigger point to enter and exit the market. It is common to calculate MA of the volumes of transactions to filter the noise and thus have a better perception of activity in the assets under study.

The MA's more used by traders are the SMA with the time periods of 200, 150 and 21 days, calculated on the close daily prices [3]. See Fig. 3.13.

The choice of the period is the most important factor in defining this indicator, since with a shorter period may incorporate an amount of noise still very high, and this will result in many more trades done by the trading system, while a longer period eliminates this inconvenience, but creates other problem, the entry point can be too late, and this mean it can lose a large part of the price movement. As such the best period choice depends more of the strategy used, the time frame of the trading system, and the financial instruments to trade.

There are two methods used in systems with MA's as trigger for the trade, the first method is the intersection between the MA and the price, in the case of open a long position, the market entry occurs when the price crosses the moving average from below to above. The second method is used when two MA crosses. The signal is given when the MA with shortest period cross from below to above the MA with longest period, and vise versa in the case of a short position.

3.6 Conclusions

In this chapter, it was described the architecture of the system and how to get the financial data used. It is presented a description of each financial ratio used, it is demonstrated why companies with good ROE show higher P/E Ratio, how to use PM ratio with increase of revenue to find the best business and to select a stock. It is explained the technical moving average indicator, and the use of it in a trading system.

References

1. D. Hillier, S. Ross, W. Randolph, J. Jeffrey, B. Jordan, Corporate Finance, Mcgraw-Hill, First European Edition ed. (2010)
2. A. Schroeder, The Snowball Warren Buffett and Business of life. Actual Editora (2007)
3. F.B.d. Matos, Ganhar Em Bolsa, Dom Quixote (2007)

Chapter 4
Multi-objective Optimization

Abstract This chapter describes the Multi-objective approach used in this work. Moreover, a complete description on the algorithm components, such as, chromosomes, genes, etc., the evaluation mechanism and the investment simulator is presented.

In Sect. 4.2 the methods for making the evolution of populations namely crossover and mutation methods are described. Section 4.3 describes the operation of an external file to retain the best individuals found in each iteration. The description of genes used in both chromosomes models that define the trading system will be presented in Sect. 4.4. The restrictions used in the investment models are explained in Sect. 4.5, the simulator is described in Sect. 4.6, and the measures for evaluation of the portfolios in Sect. 4.7.

4.1 Chromosome Structure

The representation of the chromosomes is similar for the two investments models, where each individual of the population is composed of a sequence of values called genes. The Chromosome is divided in two parts, the first group is the financial ratios weights and the second one is the trading parameters.

4.1.1 First Model of the Chromosome

This model only invests in companies that have some potential for a good return, using the first parameter of trading, the gene of *Min value for Portfolio* optimized, to filter the candidates to enter in the portfolio. This gene defines the minimum valuation of the stock that a candidate needs to have to enter in the portfolio.

A.D. Silva et al., *Portfolio Optimization Using Fundamental Indicators Based on Multi-Objective EA*, SpringerBriefs in Computational Intelligence,
DOI 10.1007/978-3-319-29392-9_4

The first seven genes are financial ratios weighs. The remaining genes define the behavior of the trading systems, in Fig. 4.1 is represented the model of the chromosome. The seven weights are the DR, ROE, PM, PER, GR, Variation of Commons Stock outstanding, and Growth in Net income. It uses five parameters of a technical trading system the stop loss, take profit to determine the exit points, the number of days to use in a SMA for trigger the entry point, and a size position to define the percentage to allocate to each stock. In Fig. 4.2 is represented the chromosome part that refers to the trading parameters. It is possible with this chromosome to have a lot of different strategies depending of the values of the genes corresponding to the parameters of trading. The system can have a more diversified portfolio, by using a lower value of position size, where its minimum value to allocate at each investment is 5 %, this allow the portfolio to be composed with a maximum number of 20 stocks.

The combination of Stop Loss and Take Profit, define the profits and loss of each investment, the time horizon of each investment, and the accuracy of the system. For example a stop loss of 20 % and a take profit of 50 %, this is a system with a long time horizon, if the evaluation and criteria to enter in the market is good, it will have an accuracy bigger than 50 %, because the noise in the market will not stop out many positions. But a system with the same take profit of 50 % and a stop out of 5 % will have lower accuracy, but a win position can compensate for a greater number of losses trades.

Other type of strategies possible is a short-Term Trading, with a high number of transactions, due to having a short Stop Loss and Take Profit, for example can have both a value of 5 %, this is translated in the market by a great number of trades, and if the accuracy is a 55 % or higher, is a system very profitable, with a low risk [1].

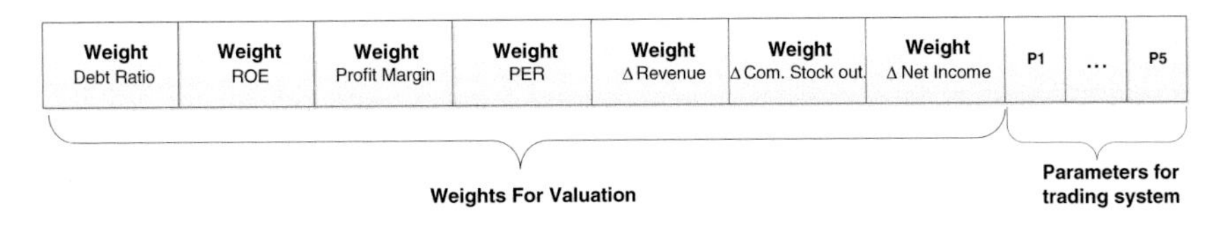

Fig. 4.1 Representation of the chromosome

Weight 1	...	Weight 7	Min value for Portfolio	Stop Loss	Take profit	Days of MA	Size of the Position

Fig. 4.2 Chromosome that represent the first model

Weight 1	...	Weight 7	Stocks Number	Size of the Position

Fig. 4.3 Chromosome representing the second model

4.1.2 Second Model of Chromosome

This model was developed to simulate a portfolio that always keeps a determined number of stocks in it. It has the first seven genes equal to the previous model, and the gene of the position size, and adds a new one, the limit number of stocks in the Portfolio.

The model operation is to maintain the stocks in the portfolio which have highest valuations, when one stock outside of the portfolio has a better valuation that the worst in the portfolio, this will be replaced by the first. In Fig. 4.3 is represented the chromosome.

4.2 Reproduction Process or Evolution Process

For doing the process of evolution in the population it was used crossover and mutation methods in each iteration, to the selected chromosomes.

4.2.1 Crossover

Crossover consists in recombining genetic parts of the chromosomes in the population selected in the mating selection, in order to create new and improved chromosomes. Next are presented the two methods used in the algorithms.

(a) **First Method of Crossover**

This method consists in choosing two chromosomes randomly from the mating pool, and recombines them to generate two new chromosomes. It is divided each chromosome into two equal parts, the first six and last six genes of the chromosome, the child's are produced by joining the set of first part of the first chromosome with the second part of the second chromosome, to create the first child, and to the second child is the inverse process of selecting the parts. In Fig. 4.4 is represented an illustration of the process.

This method was proved to be inefficient, since the population was very slow to evolve in terms of time resource, requiring a high number of iterations to generate meaningful population learning. Considering that 20 iterations took about 24 h of

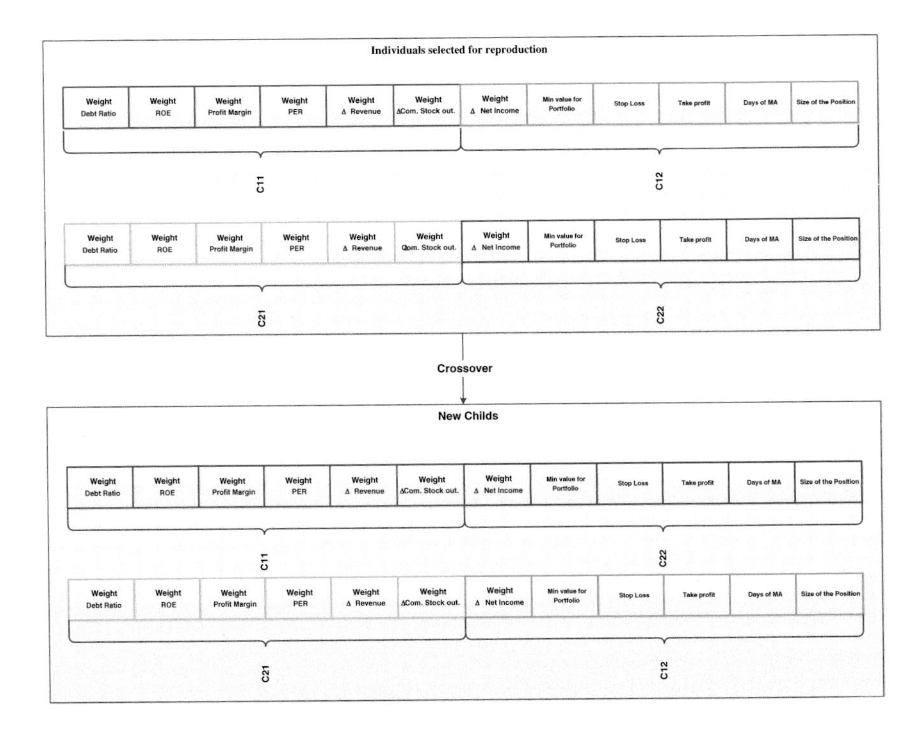

Fig. 4.4 Crossover method using two chromosomes

simulation, this method was only applied in the simulation of single objective in the first model to do a pre-training of the population.

(b) **Second Method of Crossover**

Due to problems identified in the previous method, it was necessary to improve the ratio of population learning by iteration for each simulation, so with a smaller number of iterations it is possible to obtain a better evolution. To achieve this it is selected from mating pool four chromosomes randomly, to provide their genes for a new chromosome, as illustrated in Fig. 4.5. This method has the advantage to accelerate the evolution of the population and increasing its diversity, but it is necessary to produce a sufficient number of new individuals, which means repeating the process in the same iteration more times that the method before.

4.2.2 *Mutation*

Mutation is an operation performed on individuals randomly selected of the new population generated in the crossover and from the external file. It is an operation that creates a new individual by copying the selected chromosome to the mutation,

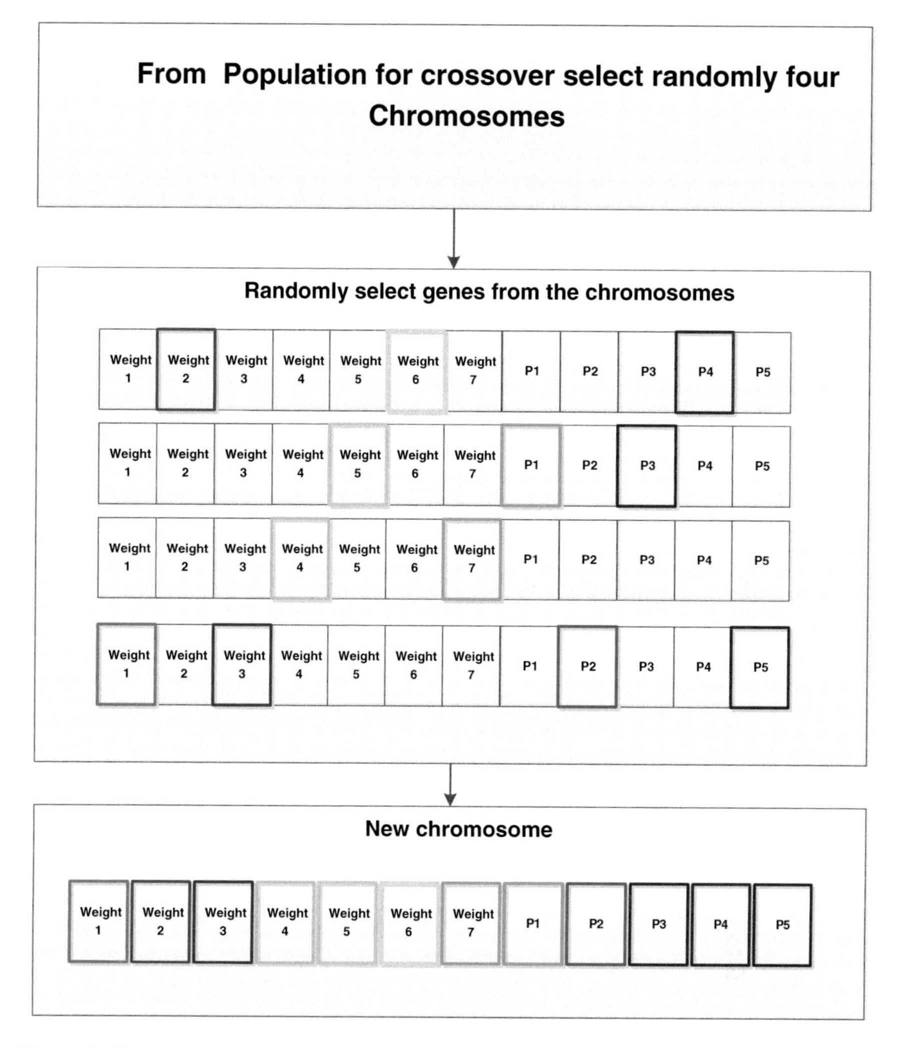

Fig. 4.5 Crossover method with four parents a one child

and modifying some genes according with a mutation rate, that is randomly and limited inside one interval.

Next it will be explaining the two methods applied in the algorithms to carry the mutation process.

(c) First Method of Mutation

This method chooses randomly four genes from the chromosome, and each one is multiplied by a random mutation rate limited between 0.7 and 1.3, as illustrated in Fig. 4.6.

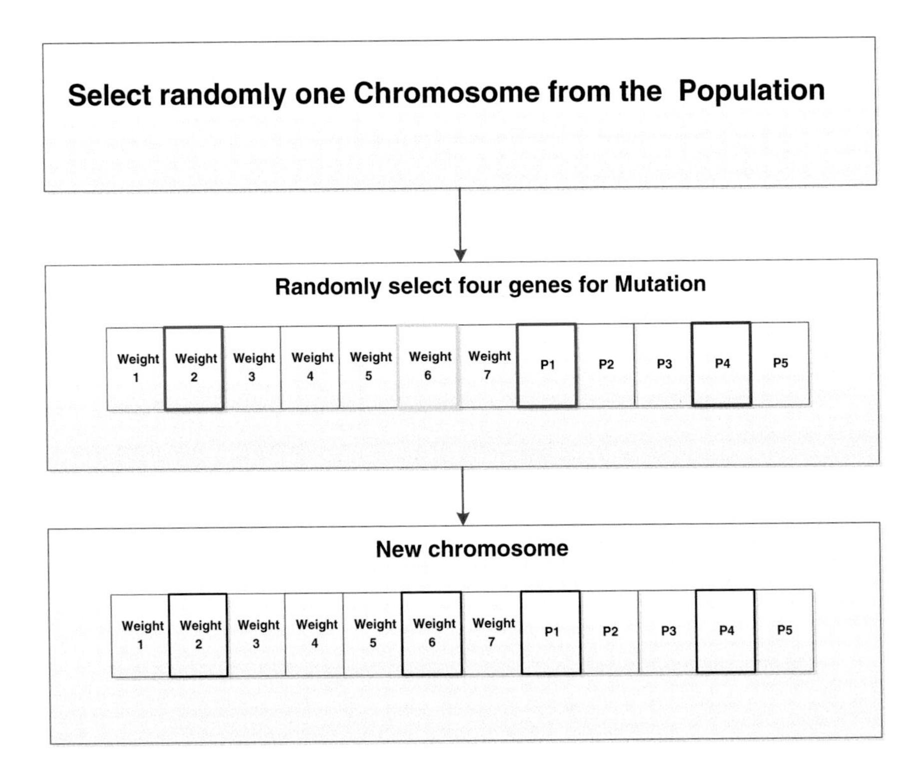

Fig. 4.6 Mutation operation with four genes mutated

(d) **Second Method**

After using the first method, it was observed that most of the mutated chromosomes, after evaluation do not have conditions to belong to the external file due to poor evaluation. Because changing four genes normally improve one and deteriorate other, this does not create a better child and the population does not evolve. Another method has been implemented which is similar to the previous one, but the number of genes to be mutated has become random, being limited between one to four genes, and the mutation rate is changed to the interval between 0.5 and 1.5.

4.3 External File

The external file holds the population composed by non-dominated solutions found along the search process, or in the case of the single objective algorithm, keep in the file the chromosomes with higher valuation of the fitness function found in the training.

(e) File Logic for Multi-objective Optimization

The function of the file is to record the individuals non-dominated and update them at each iteration of the training. To enter in the file each new individual is compared with all individuals in the file, to check if it dominates any member, the elements dominated are deleted from file and the individual is inserted, but if at least one element of the file dominates the new individual in consideration, it will not be added. In the case of an individual non-dominated by anyone in the file, it is added to the file.

In Fig. 4.7 is represented the work of the file, in the *generation I* are recorded the last population produced from *generation I-1* and the solutions that compose the pareto set found. The process starts by comparing both sets, and selects the non-dominated chromosomes to go to the reproduction process (crossover and mutation) and update the file.

(f) File Logic for Single-Objective Optimization

In case of one objective to optimize, or a function composed by two objectives like the Sharpe ratio the size of external file was limited in terms of space, for a

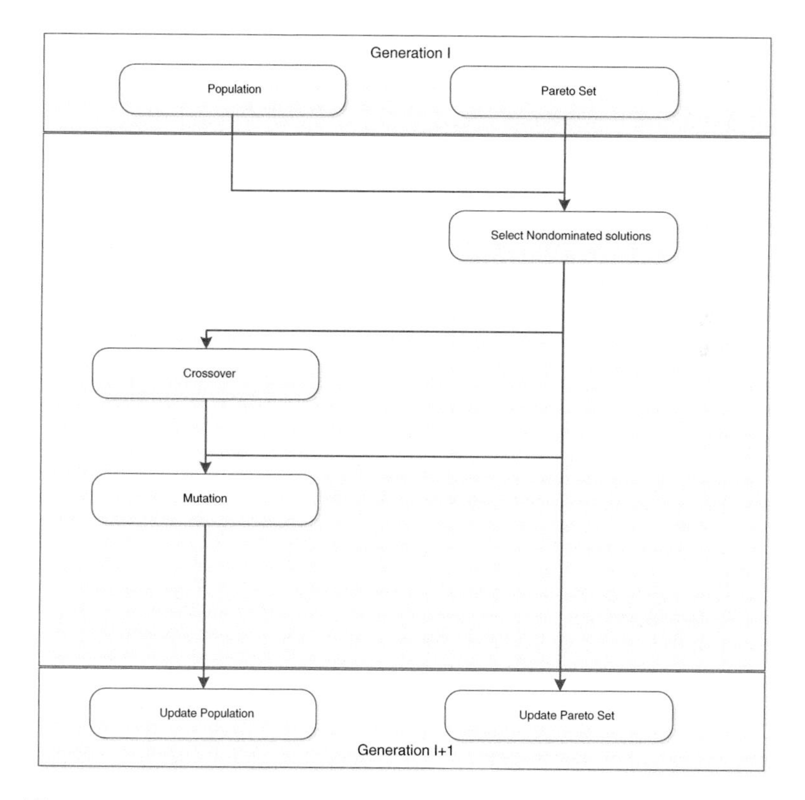

Fig. 4.7 Process diagram of the external file

determined number of solutions. Only enter in the file the best 80 individuals, and when is full if exists an individual with better fitness that the worst in the file, the worst will be deleted and the new one will be added to the file.

4.4 Parameters of the Trading System

For defining the behavior of the trading system, like the entry and exit time, the money management rules, is used the part of the chromosomes called the parameters of the trading system, for the first model is used four genes as parameters, the Stop Loss, Take Profit, Position Size, number of days of SMA, and in the second model it's used the position size and the number of stocks.

4.4.1 *Stop Loss*

The gene corresponding to the stop loss percentage, defines the maximum loss for each investment, by defining the exit point, when price tendency not occurs as expected, the stop price is calculated using the Eq. 4.1.

$$P_{\text{Stop}} = P_{\text{Entry}} \times (1 - \text{StopLoss}) \tag{4.1}$$

4.4.2 *Take Profit*

The gene corresponding to take profit has the function of deciding the percentage profit gained for all investments made. The gene will determine the sale value of successful investments based on the purchase price of the shares. The sell price is calculated by the price of entry more the value of price change, determined by the gene, according to Eq. 4.2.

$$P_{\text{profit}} = P_{\text{Entry}} \times (1 + \text{takeProfit}) \tag{4.2}$$

4.4.3 *Position Size*

The position size gene determines the percentage of the current portfolio value to invest in each new stock, this defines the level of concentration or diversification in the portfolio for each chromosome. For example a strategy that invests a low percentage in each investment, the portfolio can have a higher number of assets, if the value of position size is high, for example 20 % which is the maximum possible due to *Quantity constraint*, the Portfolio will be limited to 5 stocks. With a higher concentration, the portfolio has higher risk since it is exposed to a less number of unsystematic risks. The Eqs. 4.3 and 4.4 are the implemented calculations in the algorithms to define the capital to invest.

$$\begin{aligned} \text{If Capital}_{\text{available}} &> P_{\text{size}} * (\text{Balance}_{\text{Portfolio}} + \text{Capital}_{\text{available}}) \\ \text{Invest}_c &= P_{\text{size}} * (\text{Balance}_{\text{Portfolio}} + \text{Capital}_{\text{available}}) \end{aligned} \tag{4.3}$$

$$\begin{aligned} \text{If Capital}_{\text{available}} &< P_{\text{size}} * (\text{Balance}_{\text{Portfolio}} + \text{Capital}_{\text{available}}) \\ \text{Invest}_c &= \text{Capital}_{\text{available}} \end{aligned} \tag{4.4}$$

4.4.4 *Trigger Signal by SMA*

This is a gene used to calculate the time period of the SMA to trigger the purchase action. Sometimes the share price of the company reached a point where is undervalue, but due to market volatility or general tendency of the market, the stock price can continue to fall or get in range during some time. To achieve a better timing of enter in a long position by the algorithm is used an SMA, with the time period optimized that avoids as many as possible of the two previous scenarios and enter in the market when a Bull tendency has been establishes. The operation is that when the stock price is above the moving average, this stock may enter in the portfolio.

The SMA was not used as trigger point in crossover as is more common in systems of trading, for two reasons, first may be cases where there is no capital available to make the purchase at the time, the second reason is the portfolio is full when the price crosses the moving average, but after can be sell one asset, and the stocks can enter in the portfolio.

4.4.5 *Global Value Input for Portfolio*

The Global Value of each stock is a valuation done daily using the financial ratios, to ranking the stocks for the algorithm make trading decisions, these values are calculated using the vector of weights of ratios of the chromosome multiplied by the respective ratios of the company, as demonstrated by Eq. 4.5.

$$\text{Global}_{\text{value}} = \sum_{i=1}^{n} \text{weight}_i * \text{Ratio}_i \tag{4.5}$$

(g) Gene of the Min Value for Portfolio

The *min value for portfolio* is the gene that defines a minimum score of global value for a stock to be accepted in the portfolio, serves to ensure that only the better investments are taken when the price is reasonable.

The global value after being calculated for all firms, the companies with higher values have priority to enter in the portfolio, but only can enter if the value is higher that the value of the optimized gene. The objective of the optimization of this gene is to ensure that as the algorithm learns to choose the best companies by weighting the ratios, these companies will have better valuations.

In Fig. 4.8 is represented a flow chart of the process of valuation and order the stocks, this process is repeated for all stocks each day.

4.4.6 Stock Number in the Portfolio

This gene is only applied to the second model, and has the function to define the number of stocks in the portfolio.

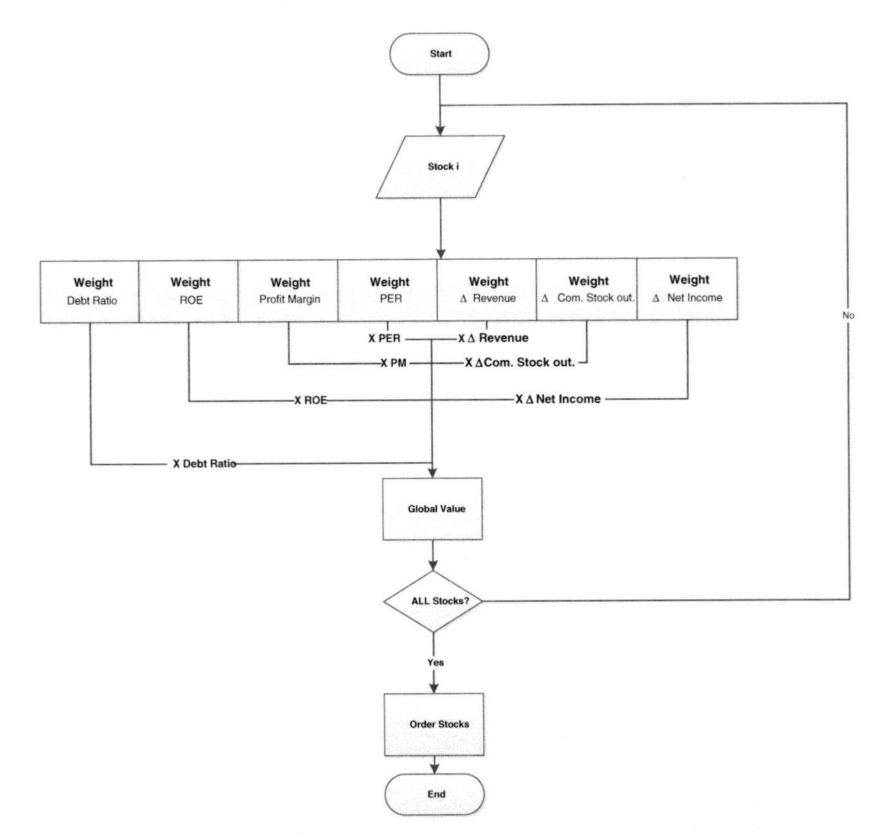

Fig. 4.8 Flow chart for evaluate the stocks

4.5 Constraints

For the simulations of the portfolios to be more realistic as possible in the design of the models was included some real constraints. For the first model it was considered the following four, while for the second model is excluded the transaction costs constraint.

Cardinality constraint limited the portfolios to a maximum of 20 stocks, the limit was chosen based on obtaining the benefits of a diversified portfolio, and studies on mathematical models that show maintaining a very diverse portfolio of stocks about 30 presents a higher level risk reduction [2].

The limit allows greater economic monitoring of the selected companies, like the philosophy of Warren Buffett, in which it should be possible to focus more attention on a smaller set of companies instead of making a great diversification, because the investor has better knowledge about the company, its products, competitors, debt levels, and future long-term economic prospects of the business [3].

Quantity constraint was used to limit the gene position size, by putting a maximum and minimum value for the size of the position. The minimum limit is set to 5 % of the portfolio value at the time of the transaction, and a maximum value is 20 %. The lower limit is to avoid positions practically insignificant to the performance of the portfolio, and the upper limit is to avoid too much exposure or weight for any stock.

Long only constraint signifies that it is not allowed to perform short selling operations, meanings the weight invested in any stock is always positive. It is a restriction used by a great number of value investors and institutional funds because the risks associated with the short selling it is considered higher than with long positions.

Transaction costs include brokerage fees, and bid-ask spreads, are costs occurred by doing the business. The transactions costs if are high, will affect the frequency of trading, by reduction the return of the portfolio [4].

In the first model is used a commission of 2 % of the value of each transaction for incorporate real costs that happen in world of investments, but for the second model, the transaction costs are ignored, to allow the algorithm change the assets without restriction, and affecting to much the return.

4.6 Simulator

The simulator in each new day uses the adjusted daily closing prices of stocks, and the financial ratios, to rank in an increasing order the stocks with higher global value, see Fig. 4.9. The block of trading, make the transactions according with the investment model used and the values of the chromosome tested. Each transaction is recorded in the balance of the account, and it is calculated daily the monetary value of the Portfolio, to calculate the variance and return of the portfolio.

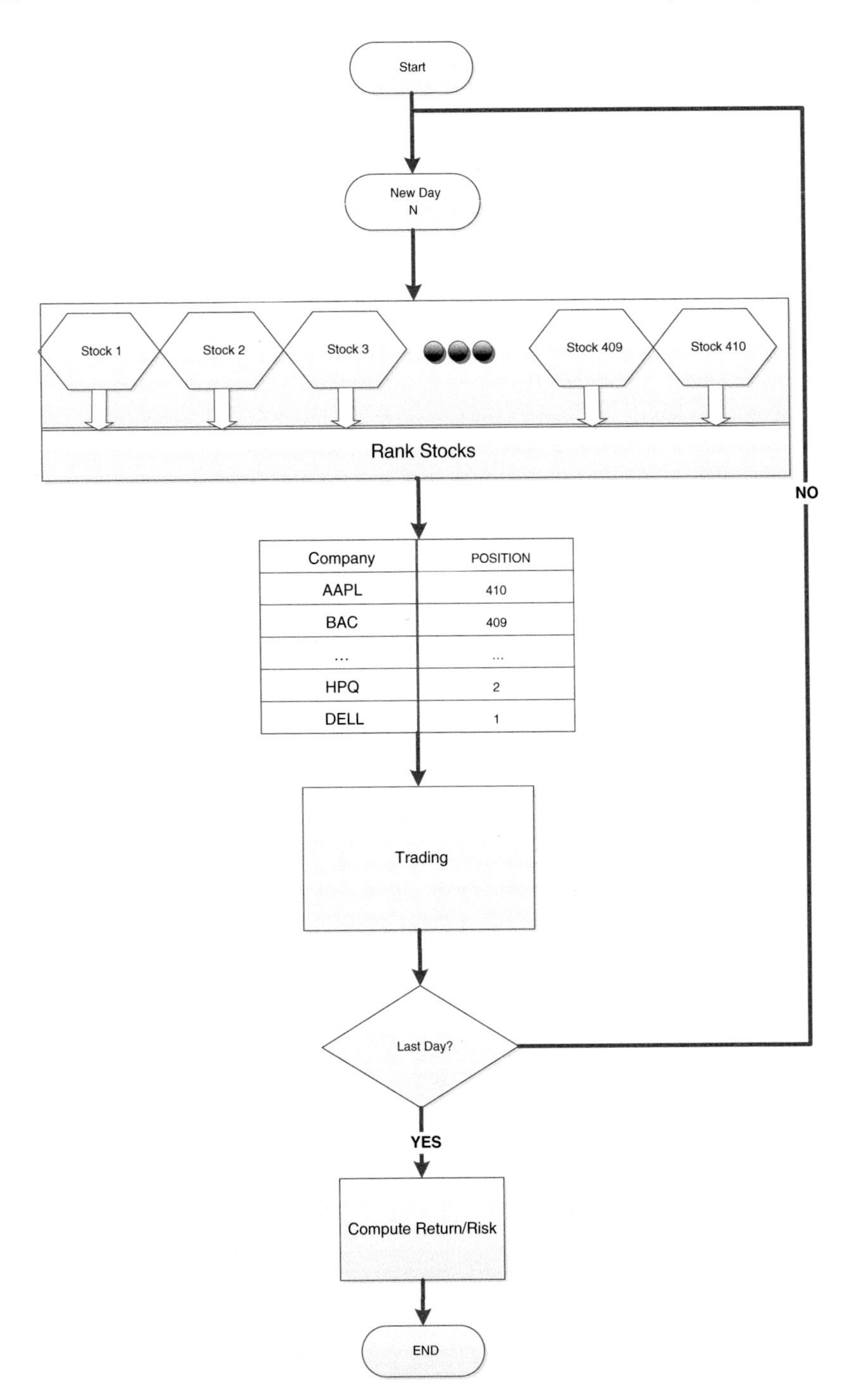

Fig. 4.9 Flow chart of the simulator of trading

4.7 Measures of Portfolio Performance

The performances measures used to do the fitness of the solutions are the ROI, Mensal Variance of ROI, and the Sharpe-ratio. For the multi-objective simulations it is used the Pareto concept, with the ROI and Variance as objectives.

4.8 Calculation of Return of Investment

The return of investment in each simulation is calculated by the sum of the value of existing stocks (calculated by the current price of the stock P_i times the number of shares N_i), and the net cash at the end of the simulation, divided by the initial capital.

$$\mathrm{ROI} = \frac{\left(\sum_{i=1}^{n} P_i * N_i\right) + \mathrm{Capital}_{\text{Last day of simulation}}}{\text{Initial Capital}} - 1 \tag{4.6}$$

4.8.1 Variance as Measure of Risk

The risk can be defined as the likelihood of deviation from the expected return of an asset in a time period. The dispersion or spread of a distribution measures how much a given return may deviate from their average, if dispersion is very high the returns that occur are very uncertain, and as such has a higher risk than a distribution with a lower dispersion, where returns are more accurate forecast.

The risk measure used is variance of the monthly returns of the portfolio that is calculated by using the Eq. 4.7 to calculate all monthly returns of the period, and uses them in Eq. 4.8 to obtain the variance.

$$\text{Mensal Net return}_i = \left(\text{Return}_i - \text{Return}_{(i-30)}\right) / \text{Return}_{(i-30)} \tag{4.7}$$

$$\sigma^2 = \frac{1}{n} \times \sum_{i=1}^{n} \left(\text{Mensal Net return}_i - \overline{\text{Mensal Net return}}\right)^2 \tag{4.8}$$

The risk definition described above can be seen in the Fig. 4.10, the distribution function of monthly returns of Apple has a greater dispersion than the S&P 500, making the predictions of the returns more accurately for the Index.

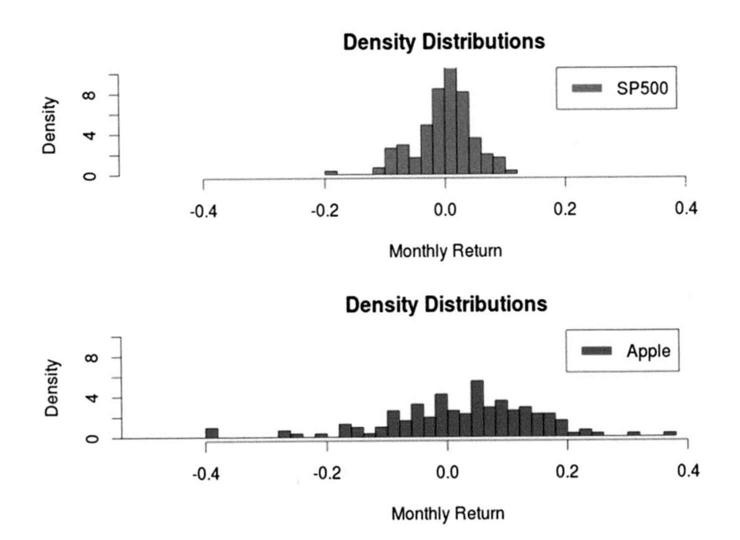

Fig. 4.10 Density function of absolute returns of Apple and Index SP&500

4.8.2 *Sharpe Ratio*

The Sharpe ratio was developed in 1966 by William F. Sharpe, is often used to evaluate the portfolios performance by pension funds and large investors. It is a simple measure of risk-adjusted performance, calculated by subtracting a risk-free rate to the return of the portfolio and dividing by the standard deviation of portfolio returns [5]. It is a reward-to-variability ratio, or return per unit of volatility, it measures the return obtained by unit of risk used.

$$\mathrm{SR} = \frac{R_{\mathrm{p}} - R_{\mathrm{f}}}{\sigma} \tag{4.9}$$

R_{p} Mean Return of the Portfolio
R_{f} Risk Free rate
σ Portfolio Standard desviation

The Sharpe ratio analyze whether a portfolio's returns are due to smart investment decisions or a result of excess risk. This measurement is very useful because although one portfolio or fund can show higher returns, it is only a better investment if those higher returns do not come with too much additional risk. A negative Sharpe ratio indicates that a risk-free asset would perform better than the portfolio being analyzed.

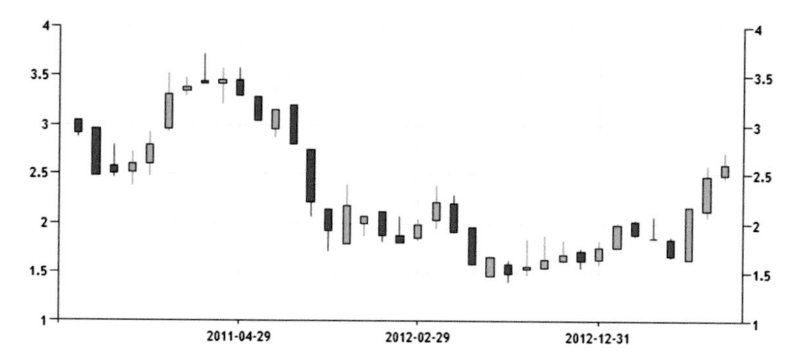

Fig. 4.11 Trend of 10 year bonds of USA

(h) **Risk Free-Rate**

The United States governmental Bond yield for 10 years observable on the graph of the Fig. 4.11 shows for the period 2011 until 07.11.2013 a yield between 1.5 and 2.5 %, these bonds are considered by the general investors as risk free asset. It was choose a constant rate of 2 % in the risk free rate for calculating the Sharpe Ratio in the Simulations.

These bonds are considered risk-free because the main rating agencies give a score of triple A to this asset, so one of the safest assets in the world. They are issued in U.S. dollars and the risk of default by the government the USA is very small, since the North American Federal Reserve can create monetary mass for the government comply with its obligations. The real risk of holding treasury bills and bonds of USA to the investors is the depreciation of the dollar that can decrease the purchasing power of the assets in USA dollars.

4.8.3 Pareto Dominance Concept

The concept of Pareto dominance ($x_1 \succ x_2$), was used in the MOEA, to determine which individuals from the population that are non-dominated with respect to objectives ROI and variance, and thus can integrate the external file. The mathematical definition for the concept is given by the Eq. 4.10.

$$\begin{aligned} x_1 \succ x_2 \quad & \text{if} \quad R_1 \geq R_2, \text{ and Risk}_1 < \text{Risk}_2 \\ & \text{OR} \quad R_1 > R_2, \text{ and Risk}_1 \leq \text{Risk}_2 \end{aligned} \tag{4.10}$$

R_i Return of Portfolio i

Risk_i Variance of Portfolio i

4.9 Conclusion

In this chapter was presented the models of investment used and the restrictions associated to them, it is demonstrated why a crossover method with random genes from four chromosomes is more reliability.

In the mutation process a bigger rate with random number of genes is better application than one with a fixed number of genes. It is explained the trading system parameters, how implement a short and long term trading system, and the measures used to evaluate the performance of the portfolios.

References

1. V.K. Tharp, *Trade to Financial Freedom* (Mcgraw-Hill, New York, 2007)
2. D. Hillier, S. Ross, W. Randolph, J. Jeffrey, B. Jordan, *Corporate Finance*, vol. First European Edition ed. (Mcgraw-Hill, New York, 2010)
3. L. Cunningham, *The essays of Warren Buffett: Lessons for corporate America* (1998)
4. G.N.A. Hassan, MultiObjective Genetic Programming for Financial Portfolio Management in Dynamic Environments. PhD Degree Thesis, University College London, 2010
5. C. Faith, *Way of the Turtle* (McGraw-Hill, New York, 2007)

Chapter 5
Results

Abstract In this chapter, the experiences with single- and multi-objective optimization will be demonstrated, and the results will be shown and the conclusion of the experiences will be explained.

This chapter is divided into three parts, the first is Sect. 5.1 where the simulations performed by the first investment model are presented, then Sect. 5.2 shows the experiences and results obtained for the second model. Section 5.3 presents the conclusions of the results obtained for the two models.

The training and real test simulations have been performed using 410 stocks of S&P 500 Index, since they had complete data for the period under study. The index is used as a benchmark for comparing the results obtained in the training and the real test. For all the simulations, the algorithm starts with an initial capital of 100,000 Dollars.

5.1 Simulations and Results of the First Model

This section presents the simulations and the optimizations for single-objective using the ROI, and the multi-objective using the Pareto concept, for the model described in Chap. 4 in Sect. 4.1.1 with the constraints of Sect. 4.5.

5.1.1 Single-Objective Using ROI

In the simulation using ROI as the objective to optimize, the population was trained during the time period from June 17, 2010 to January 3, 2012. The first training simulations performed show a tendency toward convergence of the population to a local maximum, in which the average return of the solutions found are very similar to the index return. To solve the problem, from iteration 30, the process of

A.D. Silva et al., *Portfolio Optimization Using Fundamental Indicators Based on Multi-Objective EA*, SpringerBriefs in Computational Intelligence,
DOI 10.1007/978-3-319-29392-9_5

reproduction of the algorithm is changed, from then on is used the second methods of crossover and mutation. At the end of the training, the results obtained in terms of return and variance is shown in Fig. 5.1, almost all solutions found have outperformed the index in terms of return.

(a) **Single-Objective Real Test**

The population obtained from the training was used in a real test during the period between January 4, 2012 and June 7, 2013. The results presented in Fig. 5.2 show that the returns of the chromosomes are in the same range of values as the training, thus demonstrating some adjustments between the training and the real test.

It is concluded by analyzing the results that the objective ROI used in SOEA allows to obtain a set of strategies with better returns using more risk than a passive management strategy that follows the S&P 500 index.

5.1.2 Multi-objective Using the Pareto Dominance Concept

In this experience, the objectives to optimize are the return and the variance, where it is uses the concept of Pareto dominance in the environment selection for reproduction of the population, the training was conducted in the period from June 17, 2010 to January 3, 2012.

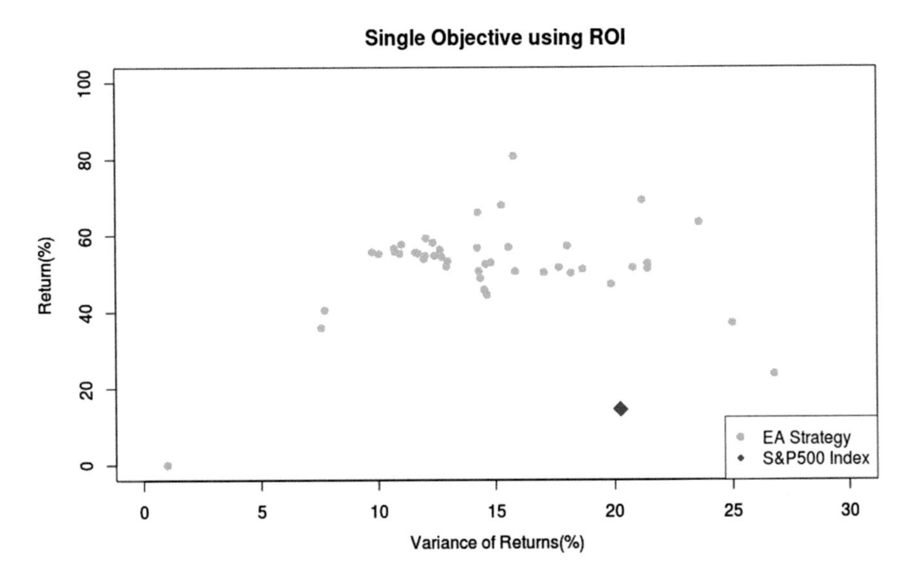

Fig. 5.1 Single-objective results obtained in training using ROI as objective

The curve in Fig. 5.3 obtained in the training has some discontinuities due to constraints, but by observing the results with the benchmark, it is concluded that the set of solutions have better results for both objectives, and a great number of solutions dominates the index.

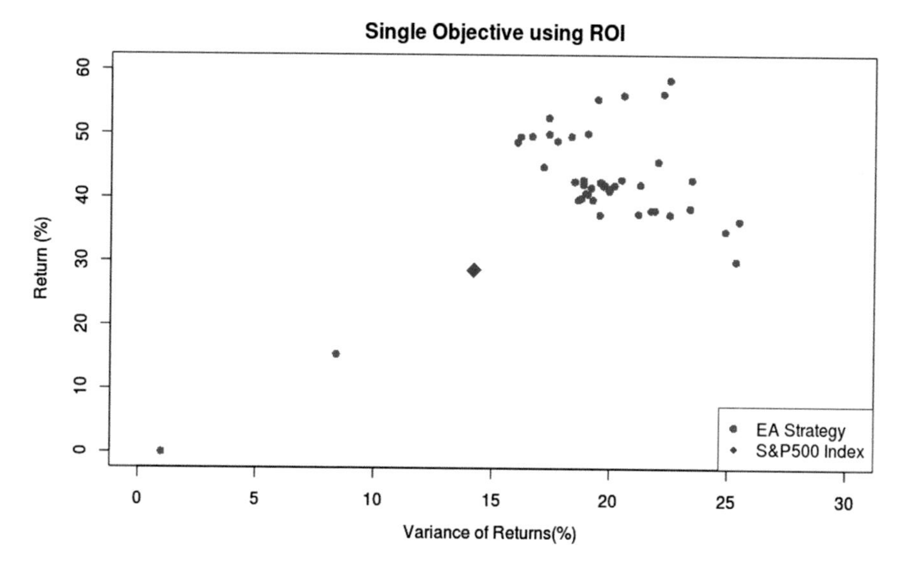

Fig. 5.2 Results of real test in single objective

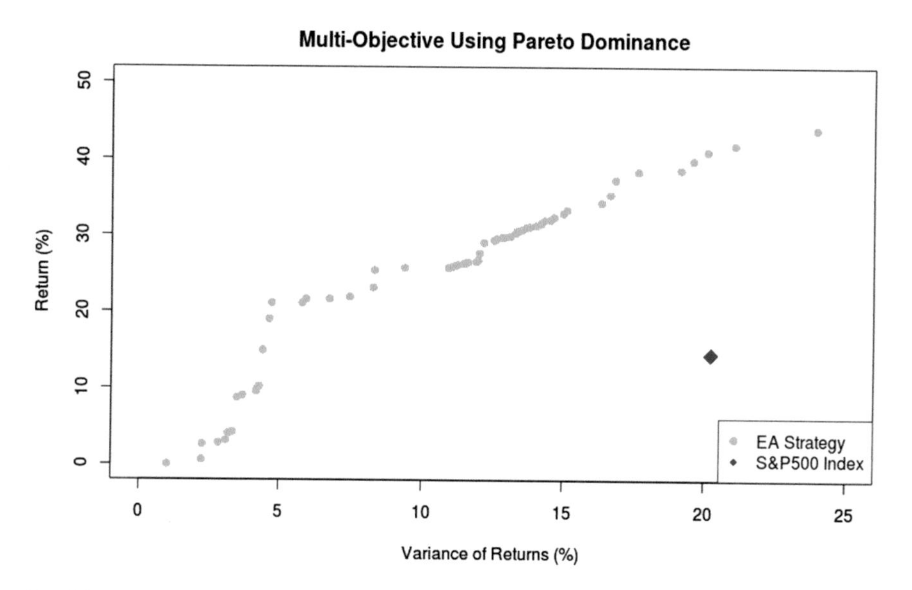

Fig. 5.3 MO results obtained in training using the Pareto dominance

(b) Multi-objective real test

In Fig. 5.4, the results of the population obtained from the training and used in a real test during the period between January 4, 2012 and June 7, 2013. These results have some differences from those obtained in the training in terms of shape of the curve, where now appear solutions that are dominated by others. The chromosomes demonstrate a capability to maintain a better performance than the benchmark, and the form of the Pareto curve is maintained.

(c) Index compared with the results

A comparison of the trend of S&P 500, and the trend obtained by a strategy that invests in all the chromosomes of the population found in the training was performed. By analyzing Fig. 5.5 it can be seen that the curve of the strategy has a better ROI, despite some effects such as over fitting in training and the transaction costs have occurred. It is possible to conclude that diversifying the capital to invest in the solutions found by the algorithm is a more viable strategy than a passive investment of buying a set of stock that are representative of the index.

(d) Analyze the Chromosomes

From this simulation five results were selected with different variances to analyze the chromosomes and to get a better insight on the type of solutions found by the algorithm. Table 5.1 shows the results for the two objectives of the chromosomes selected.

In Table 5.2 the value of the genes for the same chromosomes is presented. While analyzing the genes of the first chromosome it can be seen that it has a preference for companies with high profit margins, earnings growth, and uses little diversification. Chromosome 54 is a strategy that allows greater diversification, it allocates 5 % on

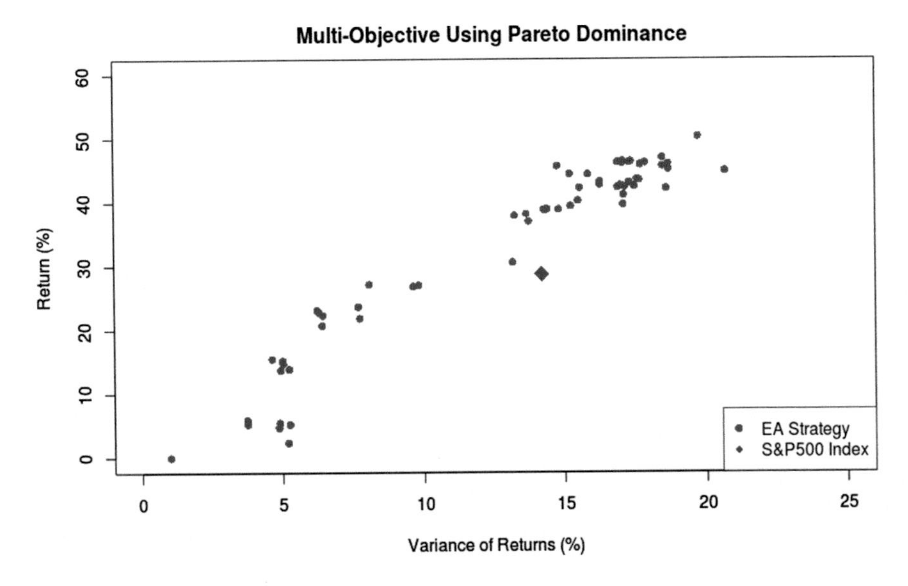

Fig. 5.4 Results obtained in real test

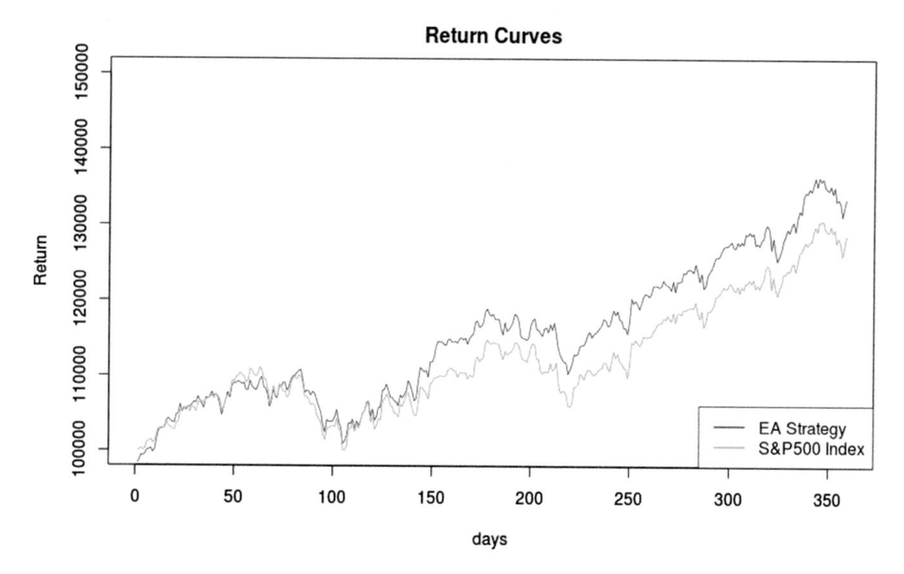

Fig. 5.5 Average return of the MOEA

Table 5.1 Results of the real test

	Return (%)	Variance (%)
S&P 500	28.685	14.20
Chromosome ID		
1	50.24	19.71
27	45.83	17.70
36	38.81	14.28
51	13.66	4.94
54	22.73	6.25

each stock, the ratios with importance are the ROE, and the net income growth rate. It is a conservative strategy, where it invests in established businesses with monopoly characteristics. Chromosome 51 is the strategy that has the higher frequency of trading, but obtained a lower return, due to the transactions costs. The analyses of the genes in the chromosome demonstrate that higher returns are achieved by a better selection of companies with a higher concentration of the investment.

5.1.3 Multi-objective Using the Pareto Dominance Concept with Two Years of Training

In this experience, the MOEA was trained with the Pareto dominance concept between June 17, 2010 and June 11, 2012. The results obtained by the solutions found are in Fig. 5.6, where there are some discontinuities in the curve, but the solutions found show some diversity along the curve.

(e) **Multi-objective real test**

The real test performed using the population obtained before, from June 12, 2012 to July 6, 2013 are in Fig. 5.7. The results had a worst performance than the simulation after. This is due to the over fitting effect, because the population was

Table 5.2 Genes of the chromosomes

Genes	Crom1	Crom27	Crom36	Crom51	Crom54
Debt ratio	0.53	0.53	0.28	0.28	0.28
ROE	0.64	0.12	2.24	0.12	2.24
Profit margin	3.01	0.61	0.49	0.96	1.01
PER	0.54	0.40	0.23	0.23	0.27
Δ of rev.	0.34	1.13	0.62	0.65	0.57
Δ common stock out	1.33	1.08	1.04	0.97	1.25
Δ net income	1.79	2.07	1.24	1.70	1.95
Global value	0.00	0.00	0.00	0.00	0.00
Stop loss	1.11	0.88	1.98	0.88	0.82
Take profit	1.89	2.46	1.18	0.12	1.36
Days of MMA	1.00	5.00	4.00	1.00	1.00
Position size	0.20	0.20	0.20	0.05	0.05

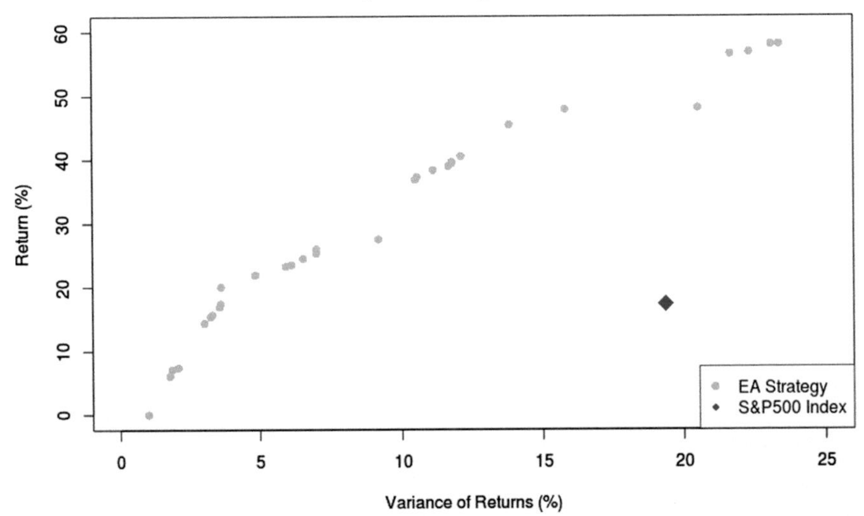

Fig. 5.6 Pareto curve obtained in training using the dominance of Pareto

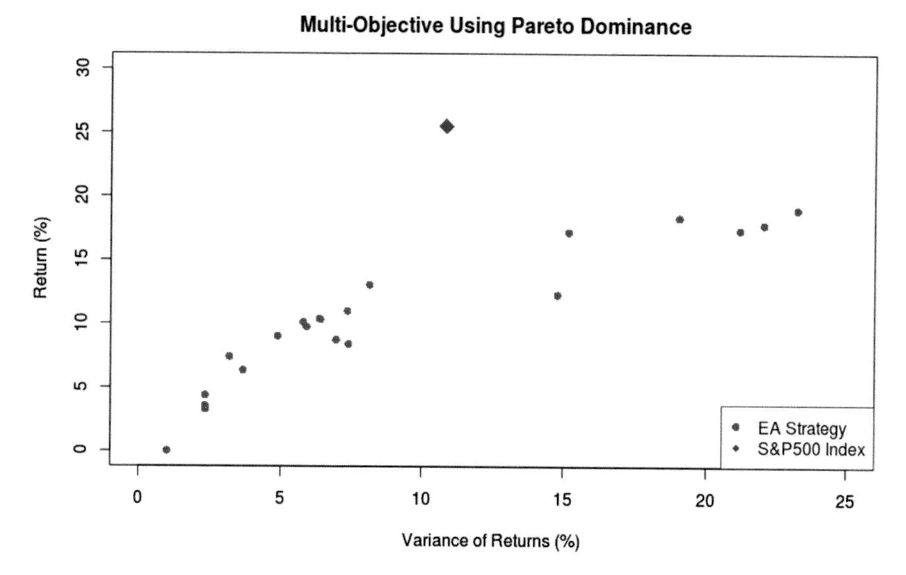

Fig. 5.7 Pareto curve obtained in real test

trained in a period where most of the stocks were undervalued, and thereby the optimization of the gene *of min value enter in portfolio* is performed with a higher value, than the *global value* for real test period of the stocks with better returns, and does not allow them to enter in the portfolio. This happens because the stocks with higher valorization, in the training period, continue in the period of real test with higher valorization than the others.

5.2 Simulations and Results of the Second Model

In this section, all the training simulations of the populations were performed from June 17, 2010 to June 11, 2012 and the real tests were executed from June 12, 2012 to June 11, 2013.

The second model is used, with cardinality constraint, quantity constraint, and long only constraint, in the next simulations. The optimizations performed are SOEA with the ROI, SOEA with the Sharpe ratio, and the MOEA with the dominance of Pareto.

5.2.1 Single-Objective Using ROI

SOEA was used for training a population with the ROI objective, the results obtained in Fig. 5.8 are in the range between 60 and 80 %, but with different degrees of risk. All solutions were found to outperform the index with a great difference in terms of return.

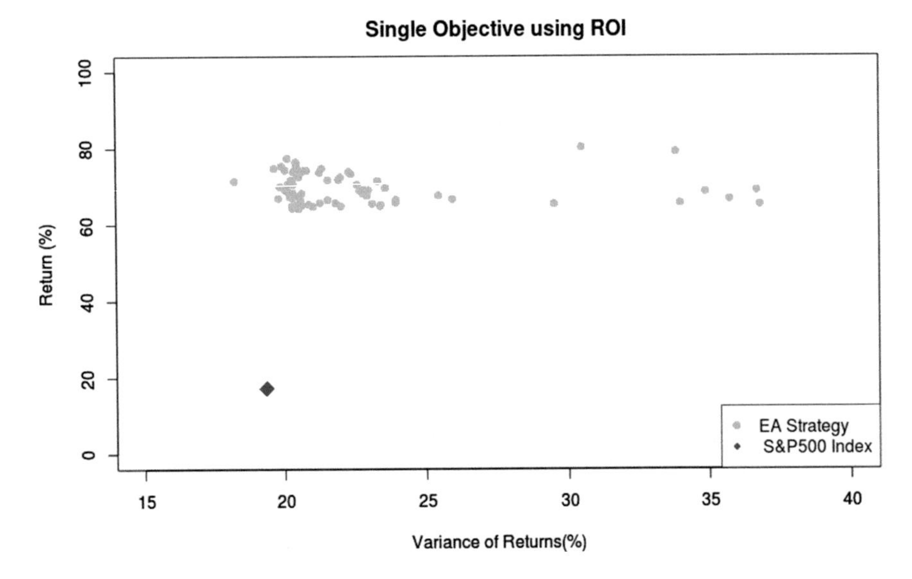

Fig. 5.8 Results obtained in training using ROI as objective

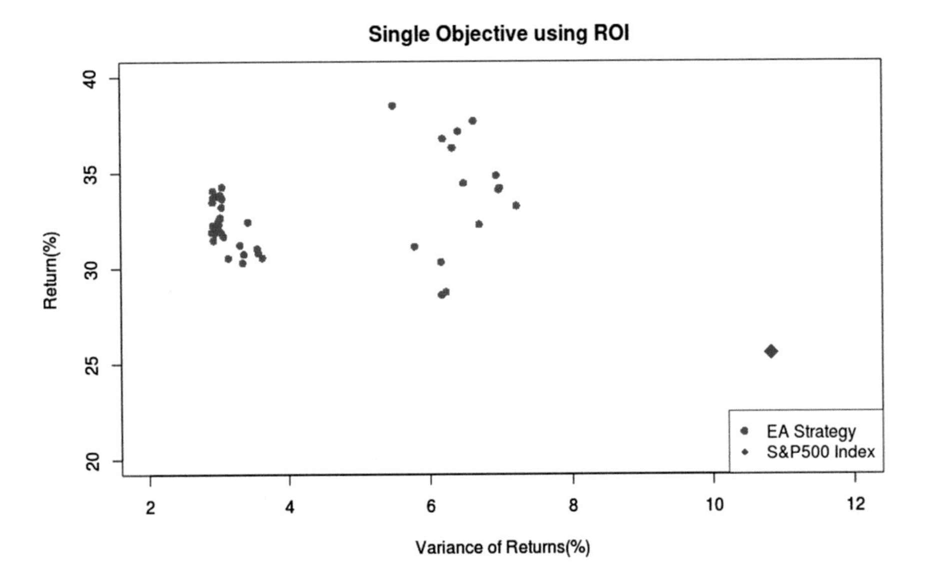

Fig. 5.9 Results of real test

(f) **Single-objective Real test**

By analyzing the Fig. 5.9, it can be observed that the chromosomes have better results in terms of return and variance than the index, this happens because the stocks selected belong to the portfolio, and in general, they have a positive tendency with higher return than the index with less oscillation in the price trend. Two accumulation zones of the results or clusters that are coincident with the results obtained in the training are defined.

(g) Analyses of the chromosomes

As previously five chromosomes were selected with similar returns but with higher difference in the variances, in order to draw conclusions about what can increase the risk of the portfolio. Table 5.3 shows the results of the two objectives of the selected chromosomes.

In Table 5.4 the value of the genes for the same chromosomes is presented. By comparing the genes, the results obtained are verified, where the differences come from the companies selected to invest that are decided by the weights to calculate the *global value*. It is verified that chromosome 29 is able to invest in companies with lower risk than the others strategies.

From this analysis, we can conclude that to construct a portfolio where the objective is to maximize the return, the important ratios in selection of the companies are the ROE, net income growth rate, and it is necessary to reduce the number of stocks in the portfolio to increase the concentration in the investments made.

Table 5.3 Results of the real test

	Return (%)	Variance (%)
SP500	25.55	10.82
Chromosome ID		
29	34.07	2.92
47	28.75	6.23
72	32.59	3.02
79	37.69	6.62
80	36.76	6.18

Table 5.4 Genes of the chromosomes

Genes	Crom29	Crom47	Crom72	Crom79	Crom80
Debt ratio	0.00145	0.00189	0.00492	0.00040	0.00229
ROE	0.14888	0.08200	0.14888	0.14888	0.10070
Profit margin	0.00002	0.00706	0.04570	0.00013	0.09422
PER	0.87995	0.87995	0.87995	0.76841	0.37111
Δ of rev.	0.00351	0.07086	0.03352	0.03352	0.05643
Δ common stock out	0.00186	0.00299	0.00911	0.00911	0.00911
Δ net income	1.49170	1.84461	1.69393	1.69390	0.80764
Stocks number	20.00000	8.00000	8.00000	13.00000	8.00000
Position size	0.20000	0.14353	0.20000	0.20000	0.20000

5.2.2 *Single-Objective Using the Sharpe Ratio as Objective*

The population in this simulation was trained using the Sharpe ratio as the objective. The results obtained in Fig. 5.10, show an increase of slope of the curve in relation to the increased variance, a behavior opposite to that obtained in relation to the Pareto curve.

(h) **Real test using Sharpe ratio**

The real test results in Figs. 5.11 and 5.12, show lower risk solutions in relation to the index, thus verifying the effect of efficiency in the use of risk by the solutions found by the algorithm while using the Sharpe ratio as an objective. By comparing this result in relation to the experience with the MOEA using the concept of Pareto dominance, it is concluded that the results of the Sharpe ratio are in the same range in terms of returns, but with less variance.

(i) **Analyzing strategies found**

From the population of the training five chromosomes were selected based on the results that they obtained in the real test, that are in Table 5.5.

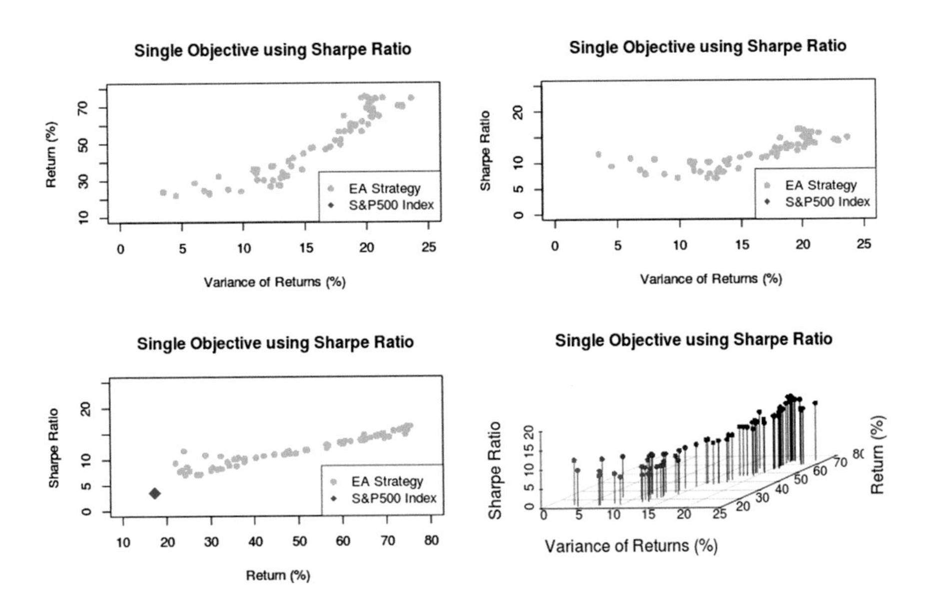

Fig. 5.10 Sharpe ratio training results

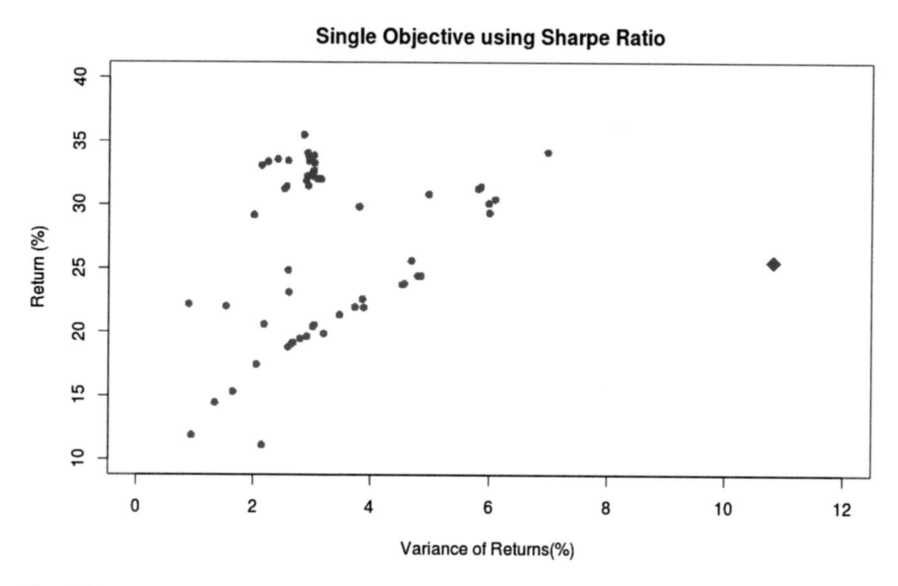

Fig. 5.11 Results obtained in the real test for the Sharpe ratio objective

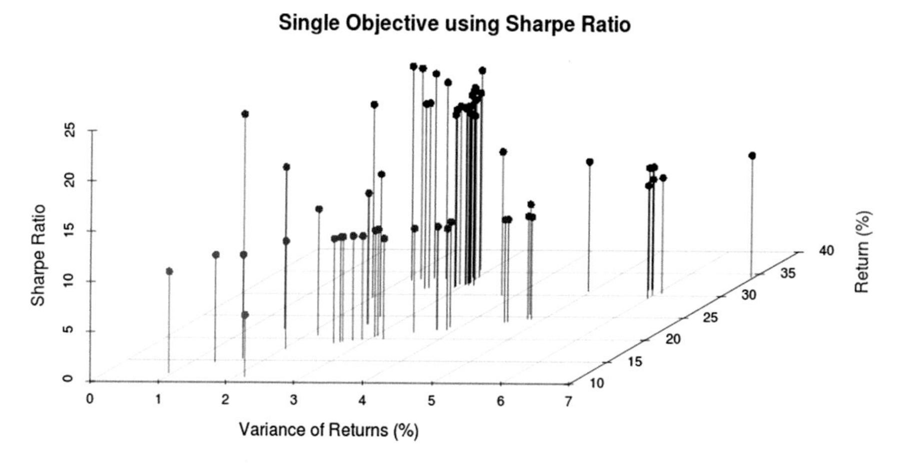

Fig. 5.12 3D graphics of real test for the Sharpe ratio objective

By analyzing the genes in Table 5.6, the difference in the results by the chromosome 7 in relation to the others are justified by more concentration of the investments, and the selection of companies with better ROE, and profit margin.

Table 5.5 Results of the real test

	Return (%)	Variance (%)
S&P 500	25.55	10.82
Chromosome ID		
7	34.00	1.71
63	23.11	2.61
68	29.19	2.00
76	17.46	2.05
79	11.88	0.95

Table 5.6 Genes of the chromosomes

Genes	Crom7	Crom63	Crom68	Crom76	Crom79
Debt ratio	0.0126327	0.005913	0.0072424	0.0072424	0.00039647
ROE	0.14888	0.002256	0.0245575	0.0150283	0.024558
Profit margin	0.0457	0.00338	0.0069581	0.0075712	0.0059008
PER	0.87995	0.8799457	0.8799457	0.4179216	0.87995
Δ of rev.	0.03352	1.0759832	0.0617327	0.03352	0.13537
Δ common stock out	0.00911	0.0200047	0.0067419	0.009113	0.010056
Δ net income	1.49167	1.5465318	0.4998159	0.8128513	1.4917
Stocks number	8	7	5	8	8
Position size	0.2	0.0892165	0.1182039	0.0741898	0.05

5.2.3 *Multi-objective Using the Pareto Dominance Concept*

In this simulation the concept of Pareto dominance in the MOEA is used, and the curve of Fig. 5.13 was obtained in the training of the population. The results show some discontinuities in the curve and little more density of solutions in the area with lower variance.

(j) **Multi-objective real test**

The population of the training obtained the following results in Fig. 5.14 in the real test. The results demonstrate a better performance than the index. Comparing the training of the MOEA and the real test curves, the results had a worst performance in real test due to the over fitting effects. It can be seen that the solutions have the ability to maintain better returns than the index for the same level of risk in both the training and real test.

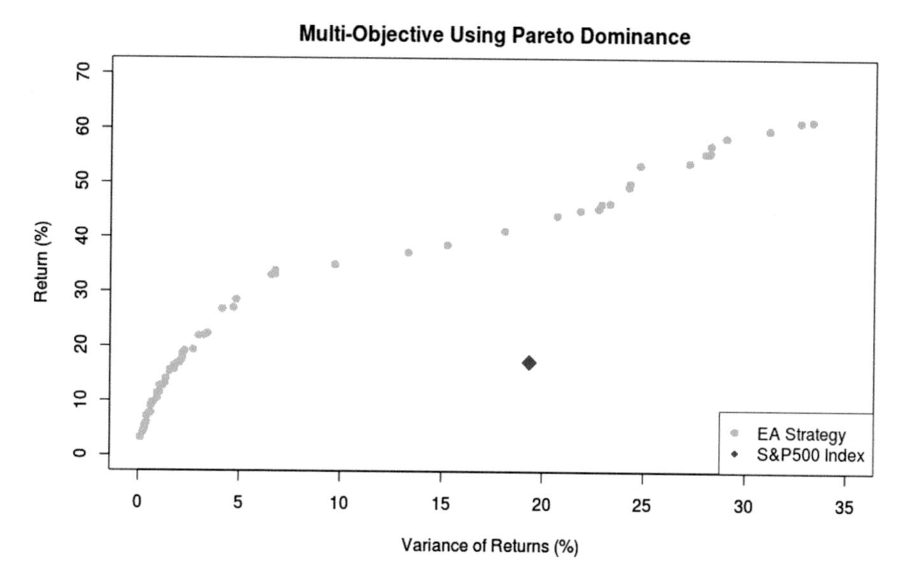

Fig. 5.13 Pareto curve obtained in training using the dominance of Pareto

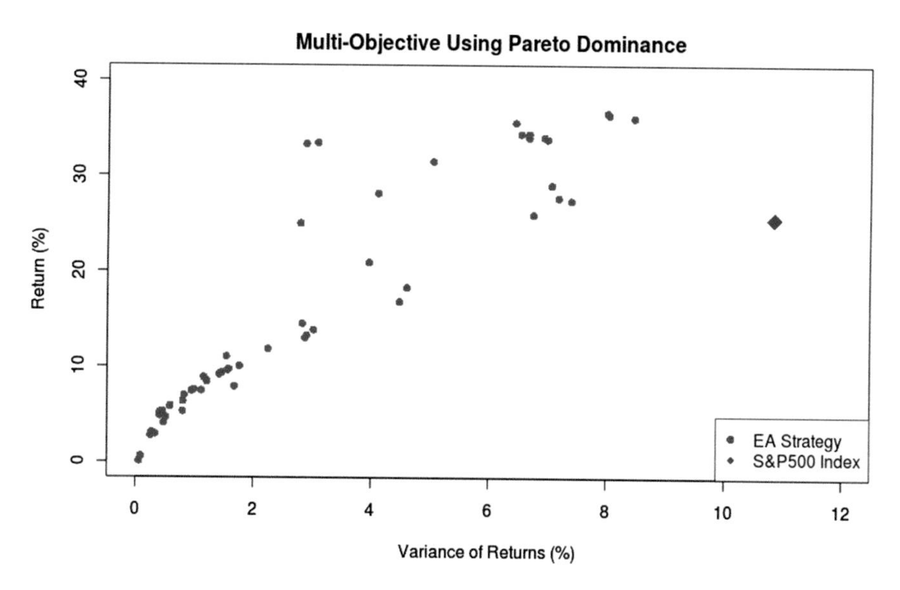

Fig. 5.14 Results obtained in the real test

Table 5.7 Results of the real test

	Return (%)	Variance (%)
S&P 500	25.55	10.82
Chromosome ID		
2	36.4	8.02
10	27.59	7.18
23	18.31	4.58
29	11.84	2.25
40	8.89	1.16

Table 5.8 Genes of the chromosomes

Genes	Crom2	Crom10	Crom23	Crom29	Crom40
Debt ratio	0.000006	0.000002	0.017	0.0003	0.000002
ROE	0.00015	0,00015	0.31	0.0029	0.31
Profit margin	0.3578	0.0059	0.02617	0.2156	0.00286
PER	1.11	1.11	2.099	0.8637	2.099
Δ of rev.	0.0739	0.0739	0.8549	1.0786	0.01089
Δ common stock out	1.3989	1.3983	1.9047	1.0967	1.5
Δ net income	2.9453	1.8516	0.331	0.043	0.638
Stocks number	8	7	9	6	4
Position size	0.18	0.18	0.063855	0.0638	0.0638

(k) **Analyzing strategies found**

From the population of the training five chromosomes were selected based on the results that they obtained in the real test, that are in Table 5.7.

By analyzing the genes in Table 5.8 the difference in the results by the chromosome 2 in relation to the others are justified by more concentration of the investments, and the selection of companies with better ROE, and profit margin.

5.3 Conclusions

The fundamental approach to stock investment using evolutionary algorithms for optimization of single-and multi-objectives demonstrates that the solutions found by the algorithms in general outperformed the S&P 500 Index.

The Sharpe ratio simulation demonstrates that it is a more efficient way to evaluate the portfolio, and using it as the fitness function in the process of reproduction leads to chromosomes with better results in efficiency terms in the use of risk. The analysis of the solutions with higher returns and with less variance in the portfolio prove that these results are achieved by a higher concentration of the investments, and better selection of the companies based on the financial ratios, and importantly, the ROE, profit margin, and net income growth rate.

The models of investment used allow a great number of trading strategies, tested in different times' frames, where the optimization finds strategies that outperform the index or have a good performance. The choices of the objectives to optimize are directly linked with the behavior of strategies obtained. If the objective is the ROI, a set of strategies with more concentration of investments and higher variances is obtained, in case of using the Pareto concept a set of solutions describing a curve with different risks of the solutions is obtained, in case of the Sharpe ratio the strategies are more efficient using the risk, and the slope of the curve increases with variance.

Chapter 6
Conclusions and Future Work

The approach of using Intelligent Computing to forecast financial markets represents a new field in financial markets investments. To develop this type of solution is not easy due to the different types of knowledge required, and the high competition in the markets.

The innovative approach implemented in this work uses financial analysis in conjunction with technical analysis, trading rules, and money management to implement active portfolio management strategies to achieves superior results than a passive strategy.

The EA implemented uses data of the stocks that compose the Index S&P 500, and in the models of investment are incorporated real constraints used in portfolios. To evaluate the efficiency of algorithms the results obtained were compared with the index.

6.1 Conclusion

The results obtained demonstrate that this approach can be applied to the management of portfolios with great results due to the optimization of the strategies and the capacity to analyze hundreds of companies in a few seconds by the algorithms.

From the simulations of different objectives, it is realized that the Sharpe ratio is better to train the populations with the goal to obtain portfolio management strategies with a more efficient use of the risk.

When using a model of investment that is always invested in the most undervalued stocks, the ratio that is more important is the PER for selecting the companies.

The ratios with more importance for select companies with a higher expected ROI and lower variance are the ROE and net income growth rate. It is concluded that to obtain higher returns it is necessary a higher concentration of investments and to use a higher risk.

A.D. Silva et al., *Portfolio Optimization Using Fundamental Indicators Based on Multi-Objective EA*, SpringerBriefs in Computational Intelligence,
DOI 10.1007/978-3-319-29392-9_6

In this investigation it is realized that this algorithm applied to real-world investing, using some human iteration to validate the investments by more in-depth analyses of the companies and conditions of the markets, can improve the results obtained.

6.2 Future Work

Studies in this work took place during the period after the crash of the North American markets in 2008, and as such the periods tested represent 3 years of Bull Market, leading to over-fitting of the solutions to this type of tendency.

In the future work, the challenge will be to test these algorithm strategies found in a bear market to check if still on average outperforms the index, thereby ensuring adjustment of the chromosomes in the most adverse environments of the market. Future improvements to implement in this approach are described next:

- An implementation of other risk measures and performance of portfolios as objectives in the EA to study what objective functions give the best results, and allow to obtain superior portfolio strategies.
- Allow short selling in conjunction with the longs positions, to the algorithms invested in the best companies in the bull market, and short, the worst in the bear market.
- Incorporate macroeconomic indicators in the algorithm to determine the general tendency of the market, as it selects the better strategies to use.
- Incorporate industrial indicators, and associate the companies that are affected by these indicators to try to extract conclusions about how to use them to find the best industries to invest and to short.
- Use adjustment measures, to avoid the effects of over-fitting.

Appendix A
Results Table

It is presented in the next table the trades performed in the real test by the MOEA of the simulation done in Sect. 5.1.2, for chromosome 36 (Table A.1).

Table A.1 Trades of chromosome 36

Crom36			
Date	Stock	Position	Amount
4-01-2012	DTV	Buy	20,414
4-01-2012	ESRX	Buy	16,255
4-01-2012	LLTC	Buy	12,922
4-01-2012	CLX	Buy	10,283
4-01-2012	YHOO	Buy	8191.6
4-01-2012	ACAS	Buy	6512.5
4-01-2012	EBAY	Buy	5178.6
4-01-2012	LSI	Buy	4128.5
4-01-2012	MRK	Buy	3305.3
4-01-2012	EOG	Buy	2562.5
4-01-2012	CBS	Buy	2102.5
4-01-2012	AES	Buy	1663.3
4-01-2012	NUE	Buy	1304.7
4-01-2012	WU	Buy	1059.6
4-01-2012	MS	Buy	838.56
4-01-2012	AKAM	Buy	671.77
4-01-2012	TEX	Buy	529.87
31-01-2013	TEX	Sell	1130.7

A.D. Silva et al., *Portfolio Optimization Using Fundamental Indicators Based on Multi-Objective EA*, SpringerBriefs in Computational Intelligence,
DOI 10.1007/978-3-319-29392-9

Appendix B
Results Table

It is presented in the next table the trades performed in the real test by the MOEA of the simulation done in Sect. 5.1.2, for chromosome 51 (Table B.1).

Table B.1 Trades of chromosome 51

Crom51			
Date	Stock	Position	Amount
2012-01-04	ESRX.csv	Buy	5082.7
2012-01-04	YHOO.csv	Buy	4835.2
2012-01-04	ACAS.csv	Buy	4590.6
2012-01-04	LSI.csv	Buy	4361.3
2012-01-04	EBAY.csv	Buy	4123.7
2012-01-04	MRK.csv	Buy	3936.6
2012-01-04	EOG.csv	Buy	3690
2012-01-04	AES.csv	Buy	3534.4
2012-01-04	DTV.csv	Buy	3358.2
2012-01-04	NUE.csv	Buy	3202.4
2012-01-04	CBS.csv	Buy	3030.9
2012-01-04	F.csv	Buy	2867.2
2012-01-04	HOG.csv	Buy	2709.6
2012-01-04	MOLX.csv	Buy	2575.5
2012-01-04	MET.csv	Buy	2466.4
2012-01-12	MET.csv	Sell	2657.1
2012-01-13	VLO.csv	Buy	5133.1
2012-01-18	LSI.csv	Sell	4720
2012-01-19	LSI.csv	Buy	5190.5
2012-01-19	F.csv	Sell	3120.5
2012-01-20	VLO.csv	Sell	5516.1
2012-01-23	ACAS.csv	Sell	5006.1

(continued)

A.D. Silva et al., *Portfolio Optimization Using Fundamental Indicators Based on Multi-Objective EA*, SpringerBriefs in Computational Intelligence,
DOI 10.1007/978-3-319-29392-9

Table B.1 (continued)

Crom51			
Date	Stock	Position	Amount
2012-01-23	MOLX.csv	Sell	2766.7
2012-01-24	VLO.csv	Buy	5209.6
2012-01-25	ACAS.csv	Buy	5216.9
2012-01-25	HOG.csv	Sell	2965.1
2012-01-26	LSI.csv	Sell	5600.5
2012-01-30	F.csv	Buy	5236.9
2012-02-01	LSI.csv	Buy	5232.5
2012-02-03	NUE.csv	Sell	3447.5
2012-02-13	EOG.csv	Sell	3968.6
2012-02-14	EOG.csv	Buy	5294.9
2012-02-16	ESRX.csv	Sell	5463
2012-02-16	AES.csv	Sell	3863.1
2012-02-17	EBAY.csv	Sell	4432.3
2012-02-21	ESRX.csv	Buy	5335.6
2012-02-21	EBAY.csv	Buy	5053.8
2012-02-21	AES.csv	Buy	4786
2012-03-07	VLO.csv	Sell	5618.9
2012-03-08	VLO.csv	Buy	5280.5
2012-03-08	CBS.csv	Buy	3272.1
2012-03-14	LSI.csv	Sell	5875.9
2012-03-20	LSI.csv	Buy	5373.4
2012-03-26	DTV.csv	Sell	3622.7
2012-04-11	SE.csv	Buy	5212.6
2012-04-17	ESRX.csv	Sell	5767.3
2012-04-19	ESRX.csv	Buy	5257.2
2012-04-19	EBAY.csv	Sell	5612.9
2012-04-25	EBAY.csv	Buy	5244.8
2012-04-26	ACAS.csv	Sell	5722.4
2012-04-27	ACAS.csv	Buy	5309.7
2012-07-09	MRK.csv	Sell	4252.8
2012-07-12	MRK.csv	Buy	5146.6
2012-07-20	EBAY.csv	Sell	5669.9
2012-07-26	EBAY.csv	Buy	5181.2
2012-08-08	ACAS.csv	Sell	5741.8
2012-08-09	ACAS.csv	Buy	5354
2012-08-27	VLO.csv	Sell	5673.5
2012-08-28	VLO.csv	Buy	5373.1
2012-09-07	EBAY.csv	Sell	5597.6
2012-09-13	EBAY.csv	Buy	5434.5

(continued)

Table B.1 (continued)

Crom51			
Date	Stock	Position	Amount
2012-10-18	MRK.csv	Sell	5579.2
2012-10-25	MRK.csv	Buy	5374.1
2012-11-19	YHOO.csv	Sell	5235.9
2012-11-21	YHOO.csv	Buy	5348.9
2012-12-13	ACAS.csv	Sell	5753.3
2012-12-17	ACAS.csv	Buy	5441.9
2013-01-02	VLO.csv	Sell	5919.6
2013-01-03	CNP.csv	Buy	5557.2
2013-01-03	MIL.csv	Buy	5275.4
2013-01-04	SO.csv	Buy	5553.3
2013-01-04	ED.csv	Buy	5286.5
2013-01-04	F.csv	Sell	5634.1
2013-01-08	LUK.csv	Buy	5546.9
2013-01-11	MIL.csv	Sell	5856.5
2013-01-15	MIL.csv	Buy	5564
2013-01-18	EOG.csv	Sell	5707.1
2013-01-24	EBAY.csv	Sell	5895.4
2013-02-11	YHOO.csv	Sell	5837.4
2013-02-12	YHOO.csv	Buy	5646.5
2013-02-12	LUK.csv	Sell	5985.3
2013-02-13	LUK.csv	Buy	5656
2013-02-14	ACAS.csv	Sell	5859.3
2013-02-15	ACAS.csv	Buy	5659
2013-03-05	CNP.csv	Sell	5988.7
2013-03-08	CNP.csv	Buy	5736.3
2013-04-02	YHOO.csv	Sell	6082.4
2013-04-04	YHOO.csv	Buy	5733.7
2013-04-16	LUK.csv	Sell	6116.6
2013-04-19	LUK.csv	Buy	5748.4
2013-04-25	ED.csv	Sell	5683.5
2013-04-29	ED.csv	Buy	5846.7
2013-04-29	CNP.csv	Sell	6194.2
2013-04-30	CNP.csv	Buy	5840.5
2013-05-08	YHOO.csv	Sell	6185.8
2013-05-10	YHOO.csv	Buy	5856.5

Printed in the United States
By Bookmasters